TECHNICAL EVALUATION REPORT

EVALUATION FINDINGS FOR TEKTON® INC. STEEL RUBBER BEARINGS

Prepared by the
Highway Innovative Technology
Evaluation Center (HITEC)

A Service Center of the Civil Engineering
Research Foundation (CERF)
CERF REPORT: HITEC 98-10
#40365
September 1998
Product 12

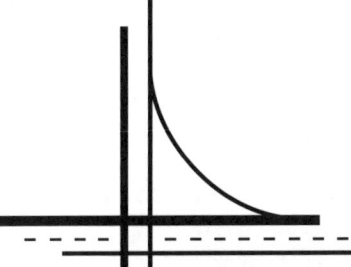

HITEC

Abstract

The Highway Innovative Technology Evaluation Center (HITEC) serves as a clearinghouse for implementing highway innovation by providing nationally-focused, collaborative evaluations of new products and technologies.

This report, *Evaluation Findings for Tekton® Inc. Steel Rubber Bearings*, presents the results of a detailed evaluation for one technology out of eleven that were evaluated in this program. The evaluations were designed to test the performance of seismic isolators and dampers produced by several manufacturers.

Library of Congress Cataloging-in-Publication Data

Evaluation findings for Tekton Inc. steel rubber bearings/prepared by the Highway Innovative Technology Evaluation Center.

    p. cm. -- (Technical evaluation report) (CERF report: HITEC 98-10)
    "July 1998."
    "A service center of the Civil Engineering Research Foundation (CERF)."
    ISBN 0-7844-0365-1
    1. Bridges--Bearings--Testing. 2. Dynamic testing. 3. Rubber bearings--Testing. 4. Seismic waves--Damping. 5. Energy dissipation. I. Highway Innovative Technology Evaluation Center (U.S.) II. Civil Engineering Research Foundation.
    III. Series. IV. Series: CERF report; 98-10.
    TG326.E86247 1998
    624'.252—dc21   98-28917
                                                      CIP

Acknowledgments

The Highway Innovative Technology Evaluation Center (HITEC), a service center of the Civil Engineering Research Foundation (CERF), prepared this report. HITEC wishes to acknowledge the special contributions of individuals whose efforts and suggestions have significantly influenced this report. Notably, this report is based on work by members of a technical evaluation panel who volunteered to develop an evaluation plan and carry out its objectives. The panel is composed of: Chairman Mohsen Sultan, P.E., California Department of Transportation (Caltrans); Steve Bradford, P.E., Alaska Department of Transportation; Maria Feng, Ph.D., University of California, Irvine; Hamid Ghasemi, Ph.D., Federal Highway Administration; Stewart Gloyd, P.E., Parsons Brinckerhoff, Quade & Douglas; Roy Imbsen, Ph.D., P.E., Imbsen & Associates; Salah Khayyat, P.E., Illinois Department of Transportation; Myint Lwin, P.E., S.E., Washington State Department of Transportation; Ayaz Malik, P.E., New York State Department of Transportation; Rolando B. Nimis, P.E., Federal Highway Administration; Walter Podolny, Jr., Ph.D., P.E., Federal Highway Administration; Charles Seim, P.E., T.Y. Lin International; Li-Hong Sheng, P.E., Caltrans; Arun Shirole, P.E. (former panel member), formerly with New York State Department of Transportation; and Edward Wasserman, P.E., Tennessee Department of Transportation. The efforts of Armand Onesto at the Energy Technology Engineering Center (ETEC) should also be noted, as he was responsible for overseeing the testing of the seismic isolators and dampers.

The writing efforts of Dr. Ghasemi, Mr. Sheng, and Mr. Onesto should also be noted as they were the persons responsible for preparing the text of all the reports developed for this project.

Caltrans was instrumental in the completion of this project as they oversaw the testing program and took the lead in analyzing the test results.

Among the staff who worked on this project, I wish to acknowledge the special efforts of HITEC's Director and CERF Vice President, J. Peter Kissinger; Kathleen Almand, P.E.; Michael S. Higgins, P.E.; Teresa Lucas; and Stacy Warner who were all instrumental to the completion of this very important project.

Publication of this report is made possible, in part, through the contributions by members of CERF's New Century Partnership:

- Black & Veatch
- CH2M Hill Ltd.
- Charles Pankow Builders
- Charles J. Pankow Matching Grant
- Kenneth A. Roe Memorial Program
- Lester B. Knight & Associates, Inc.
- Parsons Brinckerhoff, Inc.
- The Turner Corporation

Harvey M. Bernstein

Harvey M. Bernstein
President
Civil Engineering Research Foundation (CERF)

Disclaimer

Technical Evaluation Panel Key Contacts

Product: **Elastomeric Bearings**

Chair: **Mohsen Sultan, P.E.**
Senior Bridge Engineer
Office of Earthquake Engineering
Engineering Service Center
California Department of Transportation

Panelists: **Steve Bradford, P.E.**
Chief Bridge Engineer
Alaska Department of Transportation

Maria Feng, Ph.D.
Professor
University of California, Irvine

Hamid Ghasemi, Ph.D.
Research Structural Engineer
Federal Highway Administration

Stewart Gloyd, P.E.
Senior Engineering Manager
Parsons Brinckerhoff, Quade & Douglas

Roy Imbsen, Ph.D., P.E.
President
Imbsen & Associates

Salah Khayyat, P.E.
Chief, Bridge Standards and
Specifications Unit
Illinois Department of Transportation

Myint Lwin, P.E., S.E.
Bridge and Structures Engineer
Washington State Department of
Transportation

Ayaz Malik, P.E.
Associate Civil Engineer
New York State Department of
Transportation

Rolando B. Nimis, P.E.
Regional Structural Engineer
Federal Highway Administration

Walt Podolny, Jr., Ph.D., P.E.
Senior Structural Engineer
Federal Highway Administration

Charles Seim, P.E.
Senior Principal and
Senior Bridge Engineer
T.Y. Lin International

Li-Hong Sheng, P.E.
Senior Bridge Engineer
Office of Earthquake Engineering
California Department of Transportation

Arun Shirole, P.E. (former panelist)
Former Deputy Chief Engineer, Structures
New York State Department of
Transportation

Edward Wasserman, P.E.
Civil Engineering Director, Structures
Division
Tennessee Department of Transportation

Client: **Tekton® Inc.**
1565 W. University Drive
Suite 101
Tempe, AZ 85281
Phone: 602-303-9111
Fax: 602-303-9096

Paul Attaway
President

**HITEC
Project
Managers:**   **J. Peter Kissinger
Kathleen Almand, P.E.
Michael S. Higgins, P.E.**

Consultants: **Energy Technology Engineering
Center
Larry Lowe, P.E.**

Contents

Figures

Tables

Preface

The introduction of new or innovative technology to the highway community usually requires that a new product be demonstrated to many, if not all, state highway agencies. This practice is inefficient, time consuming, and often costly, particularly for small companies and entrepreneurs. To overcome these barriers, the Highway Innovative Technology Evaluation Center (HITEC) was established in 1994 in cooperation with the Federal Highway Administration (FHWA), the American Association of State Highway and Transportation Officials (AASHTO), and the Transportation Research Board (TRB). HITEC's mission is to accelerate the process of introducing technological advances to the highway community.

HITEC facilitates the conduct of consensus-based, nationally accepted performance evaluations of new or innovative technologies for the highway community. While the term "new or innovative technologies" connotes "high tech" products often associated with the computer industry, HITEC is available to evaluate almost any product, system, service, material, equipment, or other technology that the owner believes can be used on the nation's highways.

In the case of the seismic evaluation process, HITEC used a specially modified version of the HITEC process designed for group evaluations, which is illustrated below in Figure P.1.

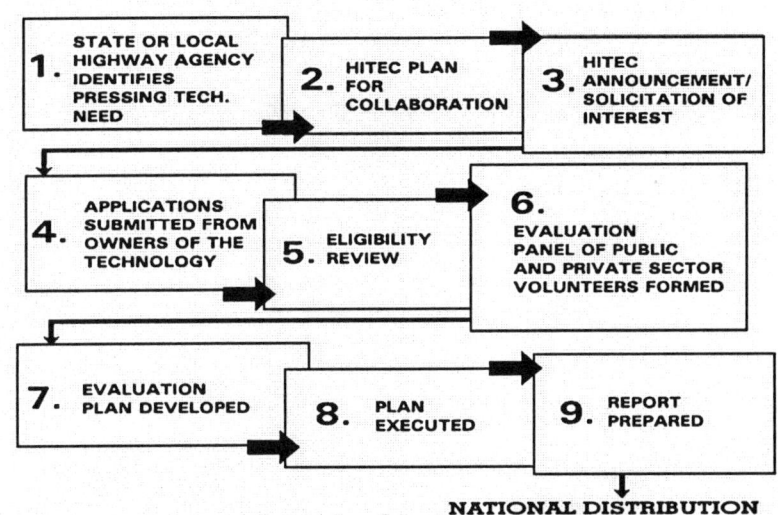

Figure P.1: HITEC Group Evaluation Process

For this evaluation, HITEC collaborated with the California Department of Transportation (Caltrans) and FHWA in the development of the Evaluation Plan. Subsequently, once the Evaluation Plan was developed (HITEC 96-02 [#40162], *Guidelines for the Testing of Seismic Isolation and Energy Dissipating Devices)*, letters were sent inviting all known manufacturers of seismic isolation and energy dissipation devices worldwide to participate in the evaluation program.

During the process of soliciting applicants to the program, HITEC staff assembled a Technical Evaluation Panel composed of representatives from the user community, academia, and the private sector. The Technical Evaluation Panel, with the cooperation and assistance of the applicants, identified the specific issues and concerns requiring resolution for these products to be adopted by the highway community. The Technical Evaluation Panel oversaw the development and execution of the Evaluation Plan.

All of the testing to implement the Plan was conducted by the Energy Technology Engineering Center (ETEC), a research testing center managed by the Rocketdyne Division of Boeing Corporation (formerly Rockwell International), which has over 30 years of experience in testing large scale components, including isolator and energy dissipating systems.

Caltrans engineering staff managed the evaluation program and took the lead in analyzing the test results and preparing the evaluation findings documented in this report. Ultimately, the HITEC Evaluation Panel reviewed and approved all of the evaluation findings.

This publication is one of 14 reports concerning the HITEC evaluation of the seismic energy and dissipation devices:

- HITEC 96-02 (#40162), *Guidelines for the Testing of Seismic Isolation and Energy Dissipating Devices* fully describes the scope and details of the HITEC evaluation program. This report is available from the American Society of Civil Engineers via 800-548-2723, 703-295-6300, or pubs@asce.org.

- This report, *Evaluation Findings for Tekton® Inc. Steel Rubber Bearings*, provides the results of the testing called for in the HITEC Evaluation Plan and measures the performance of the bearing against the criteria in the *Guidelines* report.

- Also available are 10 additional reports that provide individual results for each technology tested under the seismic evaluation process.

- A *Summary of Evaluation Findings for the Testing of Seismic Isolation and Energy Dissipating Devices* report synthesizes the performance of all units submitted for evaluation by the manufacturers that participated in the HITEC evaluation and provides basic knowledge of seismic isolation and energy dissipation.

- A *Test System Overview Report* describes the test system and methodologies used to obtain the isolator characterization data. This report describes the test equipment, test procedures, instrumentation, and data processing techniques. It also provides detailed background information of interest to engineers and technicians.

Data obtained during this evaluation is available from HITEC upon request.

CHAPTER 1

Introduction

The evaluation program for seismic isolation and energy dissipating devices was developed and executed to provide the bridge community with information and data on the dynamic performance and quality of full scale seismic isolation and energy dissipating devices designed for highway bridge applications.

Although the concept of isolation and the testing of seismic isolation devices to characterize fundamental properties is not new, the **dynamic** testing of **full scale** seismic isolation devices on a national scale has not been conducted until now.

Advances in seismic isolation technology and the potential benefits seismic isolation offers have generated much interest in recent years. However, the lack of independent, full scale dynamic characterization data and the correlation with scaled data has kept many public agencies from benefiting from what appeared to be promising technology.

In response to the heightened interest within the bridge community and the potential benefits seismic isolation offers, FHWA, Caltrans, and HITEC developed a national evaluation program for seismic isolation and energy dissipating devices.

The objectives of the evaluation program were to:

1) Implement a program of full scale dynamic testing sufficient to characterize the fundamental properties and performance characteristics of the devices evaluated;
2) Provide guidance on the selection, use, and design of seismic isolation and energy dissipating devices for different levels of performance; and
3) Help with the development of suggested guide specifications for the use of seismic isolation and energy dissipating devices in new bridges and retrofit projects.

The evaluation program examined characteristics such as:

- Range
- Capacity
- Resilience
- Performance under service and dynamic loads
- Energy dissipation

- Functionality in extreme environments
- Resistance to accelerated aging
- Predictability of response
- Fatigue and wear
- Size effects

These properties provide the bridge designer with critical information on the suitability of the devices for specific design applications and also provide insight into the reliability, longevity, and predictability of response. Furthermore, the program addressed the ability of the vendor or manufacturer to provide a quality product and predict product response.

Acceptable ranges or limits were not established for isolator or energy dissipater test performance since these requirements are typically specified on a project-by-project basis. However, target values were defined for selected parameters so that the manufacturer could provide the appropriate component for testing.

This Evaluation Findings Report summarizes and presents data collected during the HITEC Isolator and Energy Dissipater Characterization Program (HITEC Evaluation Plan) on the five steel rubber isolators submitted by Tekton® Inc. The Report describes the performance characteristics of the units that were evaluated.

This report is one of 14 reports produced in conjunction with this evaluation. Please refer to the Preface for a listing of available reports.

CHAPTER 2
Test Article Description

2.1 Description of Device

The Tekton steel rubber bearing (SRB) is a steel reinforced elastomeric isolator consisting of alternating layers of rubber and steel shims. Holes in the top and bottom steel load plates hold an array of staggered vertical tapered steel pins. In TA #1, the holes were threaded and the tapered steel pins were surrounded by sand. The sand was removed from TA #2, TA #4, and TA #5. The pins intersect holes in a steel plate located in the middle of the bearing. A neoprene cover surrounds the isolator and protects it from corrosion. During an earthquake, the rubber layers deform laterally and steel pins yield. The device is designed to reduce earthquake force on the structure by allowing the structure to translate laterally through shear deformation of the rubber layers and by absorbing energy when the steel pins deform plastically.

Tekton manufactured and submitted five isolators for evaluation. Figure 2.1 consists of a photograph of one of the seismic isolators submitted by Tekton for evaluation. Figures 2.2 shows a dimensional drawing of the 150 kip isolator.

Figure 2.1 Tekton Steel Rubber Bearing Seismic Isolator

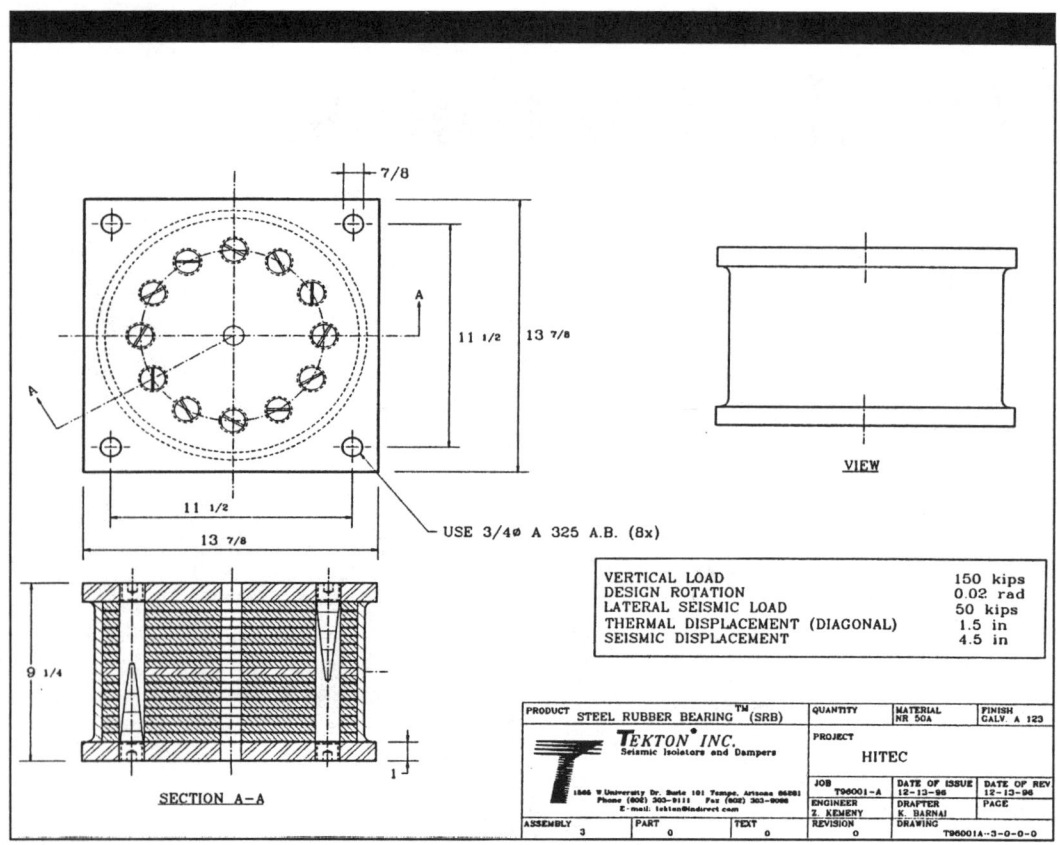

Figure 2.2 Tekton 150 kip Steel Rubber Bearing Drawing

Table 2.1   Test Article Physical Properties

Test Article ID	Design Compressive Load (DCL)	Design Displacement (DD)	Movement Rating (MR)	Weight	Height*	Elasomeric Plan Dimensions*
	(kips)	(in)	(in)	(lb)	(in)	(in)
TA #1	150	6.0	2.0 (+/- 1.0)	187	7.25	13 dia.
TA #2	500	6.75	3.0 (+/- 1.5)	700	11.5	22 dia.
TA #3	500	6.75	3.0 (+/- 1.5)	700	11.5	22 dia.
TA #4	500	6.75	3.0 (+/- 1.5)	700	11.5	22 dia.
TA #5	750	10.0	4.0 (+/- 2.0)	1,039	14.25	26.5 dia.

* See shop drawings.

Table 2.1 summarizes the Tekton SRB design parameters and physical characteristics of the five isolators submitted for evaluation.

2.2 Manufacturer's Data Package

The Evaluation Panel requested that each manufacturer submit a data package containing test results, working drawings, and any other data deemed necessary to assess the suitability of the manufacturing process. Manufacturers were required to provide this documentation prior to evaluation. Submitted information included:

■ WORKING DRAWINGS representing the test article (TA) and the device that the manufacturer intended to market showing all manufacturing details, dimensions, and allowable tolerances.

- MATERIAL SPECIFICATIONS AND DATA for all components/parts with accompanying American Society for Testing and Materials (ASTM) specifications necessary to identify the test articles for future applications. (Companies that supplied test articles manufactured to foreign standards were required to submit the closest ASTM equivalent specifications.)
- MATERIAL CERTIFICATIONS confirming that the materials used in manufacturing the test articles adhere to the material specifications.
- PERFORMANCE PREDICTIONS and supporting design calculations.
- ENVIRONMENTAL TEST DATA concerning each device's resistance to aging, ultraviolet light, ozone, salt spray, moisture, sand, and dust.
- CERTIFICATE OF COMPLIANCE confirming that the test article conforms to the submitted working drawings,

material specifications, allowable manufacturing tolerances, and the quality control plan. (The certificate should be supported by copies of test results performed on the devices and materials test reports for the component materials.)
- NAME(S) AND LOCATION(S) OF THE MANUFACTURER.
- QUALITY CONTROL PROGRAM SUMMARY describing the quality control procedures employed during the manufacturing process.

Table 2.2 summarizes the actual data package received from Tekton.

Manufacturers were required to review ETEC's test matrix prior to testing.

Table 2.2 Tekton Steel Rubber Bearing Submittal Summary

Description	Submitted Prior to Testing (Y/N)*	Submitted After Testing (Y/N)*	Comments
Working Drawings:			
Layout	Y	-	
Assembly	Y	-	
Details	Y	-	
Allowable Tolerances	Y	-	
Parts List	Y	-	
Adapters	Y	-	
Materials Specifications	Y	-	
Materials Certifications	Y	-	
Performance Predictions:			
Response Equations	Y	-	
Design Calculations	Y	-	
Pretest Predictions:			
Stiffness	Y	-	
Damping	Y	-	
Hysteresis	Y	-	
Test Data	Y	-	
Test Reports	N	N	
Environmental Data	Y	-	
Certificates of Compliance	N	Y	
Name and Location of the Manufacturer	Y	-	
Q/C Program Summary	Y	-	
Y = Yes; N = No			

CHAPTER 3

Test Setup and Test Plan

3.1 Installation

Adapters

The manufacturer provided sketches of the adapters and spacers. The drawings conformed to all HITEC/ETEC requirements. The test articles were fabricated and shipped by Tekton.

Handling Requirements

The manufacturer did not specify special handling requirements for the test articles. Threaded holes for inserting eye-bolts were provided for handling.

Installation Requirements

A locking device was provided with the test articles to prevent unwanted motion during installation. The mechanism was removed prior to testing.

Installation Problems

The adapters and test articles were installed without incident.

3.2 Performance Testing

Testing of the Tekton SRB isolators was conducted in accordance with the requirements set forth in the HITEC Evaluation Plan, which consists of nine separate tests. Each test provided a means for evaluating a specific characteristic of the test item(s):

Test 1 – Performance Benchmark: To verify experimentally the initial stiffness, damping, and number of loading cycles required to stabilize response.

Test 2 – Compressive Load Dependent Characterization: To quantify experimentally the effects of varying compressive loads on the performance characteristics, specifically stiffness, damping, and energy dissipation per cycle (EDC).

Test 3 – Frequency Dependent Characterization: To determine experimentally dynamic performance characteristics at varying frequencies in the primary direction of operation.

Test 4 – Frequency Dependent Characteristics: To determine dynamic performance characteristics and verify the constitutive laws. This test is only applicable to dampers; therefore, it was not used as part of the evaluation of Tekton's SRB elastomeric bearings.

Test 5 – Fatigue and Wear: To evaluate the potential seismic performance changes resulting from 10,000 cycles of service movements (temperature and live load fluctuations).

Test 6 – Environmental Aging: To verify experimentally seismic performance after exposing the device to a salt spray environment.

Test 7 – Dynamic Performance Characteristics at Temperature Extremes: To assess the effects of extreme temperature on the performance characteristics, namely stiffness, damping, and EDC.

Test 8 – Durability: To assess component durability resulting from a moderate number of strong motion cycles.

Test 9 – Ultimate Performance: To determine experimentally ultimate displacement and margins of safety.

ETEC performed the applicable tests on each of the test articles submitted by Tekton. The test matrix specified by the HITEC Evaluation Plan for elastomeric isolators is summarized in Table 3.1. Test details are described in the following sections.

A technical representative from Tekton was on site when early tests were preformed to observe and provide guidance for the installation of the units.

Table 3.1 HITEC Program Elastomeric Isolator Test Matrix

TA #1 (150 kip)	Test 1	Test 3	Test 2	Test 8	Test 9	N/A
TA #2 (500 kip)	Test 1	Test 3	Test 2	Test 7	Test 8	Test 9
TA #3 (500 kip)	Test 1	Test 3	Test 2	Test 7*	Test 8	Test 9
TA #4 (500 kip)	Test 1	Test 3	Test 5	Test 6	Test 3	Test 9
TA #5 (750 kip)	Test 1	Test 3	Test 2	Test 8	Test 9	N/A

Note: Test numbers do not correspond to the testing sequence. Table 3.1 shows the actual test sequence, left to right, for each test article submitted.

* Test 7 (for hot temperatures) was performed on TA #3 only if it was fabricated from different materials than TA #2.

CHAPTER 4

Data Summary, Analysis, and Review

This chapter summarizes results for each test performed on the Tekton SRB isolators. Test results and specific details regarding the individual component tests are discussed. AASHTO definitions for effective stiffness and equivalent damping ratios as defined in the *1997 AASHTO Guide Specifications for Seismic Isolation Design* manual are used. Please refer to Appendix B for additional information regarding these stiffness and damping calculations. Results from each test are reported in the following order:

1) Purpose
2) Test Procedure
3) Test Variances
4) Data Summary
5) Test Observations

Test results and normalized test results were plotted for Tests 1, 2, 3, and 8. Normalized plots were used to directly compare isolators of varying sizes. Dynamic performance characteristics such as force degradation, stiffness, damping, and EDC for various size isolators when normalized can show performance trends and variations. Variations in performance may be attributed to manufacturing processes, material variations, size effects, or other factors.

Plots were normalized on the second cycle with 70 percent design compressive load (DCL) and a 2.0 second period. The second cycle was selected to reduce test setup dynamic effects. The 2.0 second period was selected because it is the closest period to a typical isolated structure period.

4.1 Test 1 – Performance Benchmark

Purpose
To determine experimentally the initial stiffness, damping, and number of loading cycles required to stabilize cyclic response. This number of cycles is referred to as the number of shake-down cycles.

Procedure
ETEC applied the DCL and then applied 10 fully reversed loading cycles at the design displacement (DD) at a frequency corresponding to a 2.0 second period. If the device did not stabilize, the manufacturer was notified that no further testing would be undertaken.

$$F_i = \text{Peak lateral load for the } i^{th} \text{ cycle}$$

The factor F_2/F_{10} was used to indicate whether shake-down occurs. Currently, there are no definite limits to determine whether shake-down has occurred. However, based on a consensus of the Technical Evaluation Panel for this testing program, it was established that shake-down occurs if $0.7 < F_2/F_{10} < 1.3$, as documented in the HITEC Evaluation Plan. The Panel later decided that F_3 is more appropriate than F_2. Therefore, Table 4.1 uses the third cycle rather than the second cycle. The table also presents the values of EDC_3/EDC_1 and EDC_{10}/EDC_1 for informational purposes.

The resulting data was used to determine shake-down characteristics of the test article and confirm that minimum damping requirements were satisfied for the devices or combination of devices evaluated.

Test Variances

Test variances were necessary while testing Tekton's SRB devices. Each variance was documented and is presented in this report.

Variance #1: The Performance Benchmark test was not used to terminate the evaluation if a test article did not meet the evaluation protocol shake-down criteria.

Reason: The HITEC Panel deemed the requirements as described above and in the Evaluation Plan to be too stringent.

Impact: Test 2 through Test 9 were executed. Additional information on bearing performance was obtained.

Variance #2: Instrumentation cables were crossed when the accelerometer was moved from the test rig to the actuator for the testing of TA #5.

Reason: Cables were accidentally crossed by a technician.

Impact: The impact was minor and the amount of data lost was not critical.

Resolution: The system was rewired when a new accelerometer was installed.

Data Summary

The results from Test 1, which was performed to characterize test article response stability, are summarized in Table 4.1. Evaluation results are for circular bearings and are not representative of rectangular bearings.

Test Observations

The first 10 shake-down cycles for TA #1 indicated unstable performance with a significant, excessive compressive set (0.25 inches) and extensive distortion of the outer cover. This was classified as a failure and no further tests were performed on TA #1.

The unstable performance of TA #1 prompted the manufacturer to request a suspension of further testing. The suspension was granted. TA #3, which had not been tested, was shipped back to the manufacturer for modification. The manufacturer also modified TA #2, TA #4, and TA #5. The sand was removed from the cavity surrounding the steel pins on these three test articles, and the design displacements were reduced from nine inches to 6.75 inches (500 kip isolators) and from 12 inches to 10 inches (750 kip isolator). After the modifications, TAs #2, #3, #4, and #5 were tested.

The rate of degradation for TA #2 and TA #4 was rapid and unstable to the sixth cycle. A 54 percent drop in lateral force

Table 4.1 Test 1 — Performance Benchmark

Test Article ID	F_3/F_1*	F_{10}/F_3	K_3/K_1	K_{10}/K_3	EDC_3/EDC_1	EDC_{10}/EDC_1
TA #1 (150 kip)	1.01	0.54	0.92	0.56	1.00	0.32
TA #2 (500 kip)	0.91	0.52	0.88	0.51	0.95	0.13
TA #3 (500 kip)	0.73	0.71	0.73	0.71	0.53	0.22
TA #4 (500 kip)	0.87	0.56	0.85	0.54	0.87	0.13
TA #5 (750 kip)	0.80	0.56	0.77	0.60	0.78	0.18

F_i = Peak lateral force during i^{th} cycle.
K_i = Effective stiffness during i^{th} cycle.
EDC_i = Energy dissipation during i^{th} cycle.

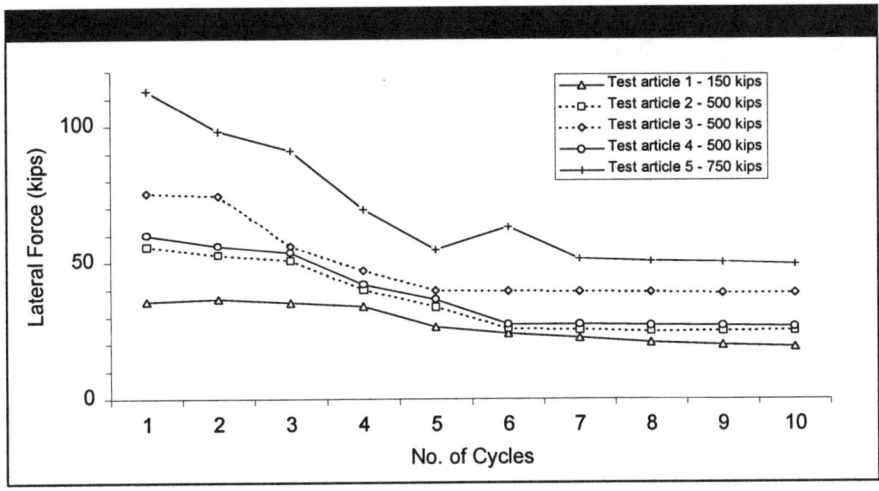

Figure 4.1(a) Force Stability During Performance Benchmark Test: Force Degradation

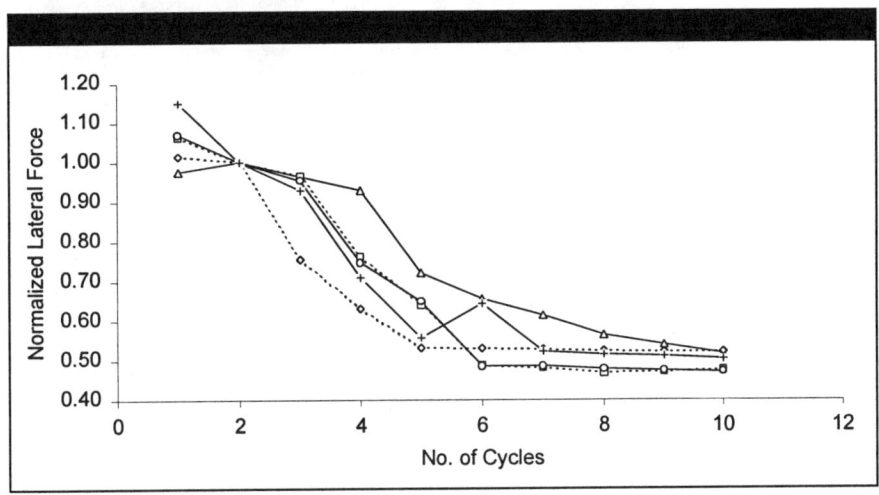

Figure 4.1(b) Force Stability During Performance Benchmark Test: Normalized Force Degradation

was recorded during the first six cycles for both test articles. The rate of degradation for TA #3 (47 percent drop) and TA #5 (51 percent drop) was rapid and unstable to the fifth cycle.

A delamination occurred on TA #2 between the second rubber laminate and the third steel shim during the fourth cycle. During the cyclic testing, the steel shims in TA #2 were observed to bend with each successive stroke. Three separate delaminations were observed between the steel shims and rubber laminates near the top and bottom load plates on TA #4. Two delaminations were also observed between the rubber laminates and the top and bottom load plates on TA #5. TA #1 and TA #3 had outer protective covers that prevented visual inspection of the steel shims and rubber laminates during testing.

4.2 Test 2 – Compressive Load Dependent Characterization

Purpose

To quantify experimentally the effects of varying loads on the performance characteristics, specifically, stiffness, damping, and EDC.

Procedure

Load deflection and damping characteristics were determined as follows. Circular devices were loaded in one arbitrarily selected primary direction of operation and at 90 degrees to this direction. Square or rectangular bisymmetric devices were loaded in the primary direction of operation, at 90 degrees to the primary direction, and in a diagonal direction through a point of the bearing surface that was farthest from its centroid, i.e., at 45 degrees to the primary loading direction for a square cross-section. The tests consisted of three cycles of fully reversed design displacement applied at a frequency corresponding to a 2.0 second period for each device. Three tests were performed in the primary direction of operation; the first at 40 percent DCL, the second at 70 percent DCL, and the third at 100 percent DCL. Tests in the other directions, when required, were performed with 100 percent DCL. If the device was unidirectional, it was tested only in that direction.

Test Variances

No variances were reported.

Data Summary

The results from Test 2 were used to assess the effects of compressive loading on stiffness, EDC, and damping of weight-bearing devices under load. These results are summarized in Table 4.2 and Figures 4.2 (a-d).

Test Observations

Compressive load dependency trends were not obvious from the test data. The stiffness decreased with increasing DCL for

Table 4.2 Test 2 — Compressive Load Dependent Characterization at DD and 2.0 Second Period (Second Cycle)

Test Article ID	Compressive Load	Stiffness, K_{eff} (kips/in)	Damping (% Critical)	EDC (in-kips)
TA #1 (150 kip)	100% DCL	**	**	**
	70% DCL			
	40% DCL			
TA #2 (500 kip)	100% DCL	3.2 [4.5]	9.2 [11.0]	80.7 [135.3]
	70% DCL	3.7	6.7	67.9
	40% DCL	4.0	7.1	77.5
TA #3 (500 kip)	100% DCL	5.6 [6.4]	14.3 [14.9]	219.7 [253.1]
	70% DCL	5.2	12.2	173.9
	40% DCL	4.7	9.9	126.8
TA #5 (750 kip)	100% DCL	4.8 [6.0]	14.2 [14.6]	410.1 [527.8]
	70% DCL	5.0	11.1	335.8
	40% DCL	4.9	8.9	261.7

[ ] Rotated 90° from initial/primary orientation direction.
** Test 2 was not performed on TA #1 due to failure during Test 1.

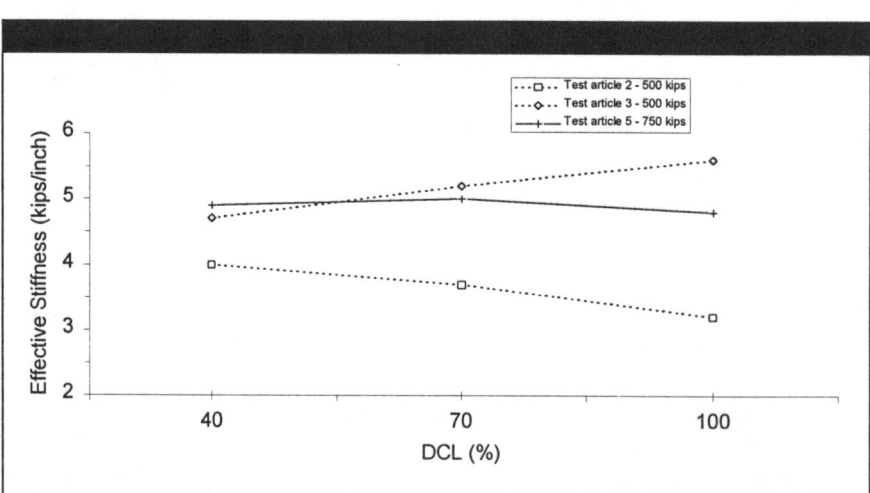

Figure 4.2(a) Compressive Load Dependency: Stiffness Compressive Load Dependency

HITEC: Tekton® Steel Rubber Bearings

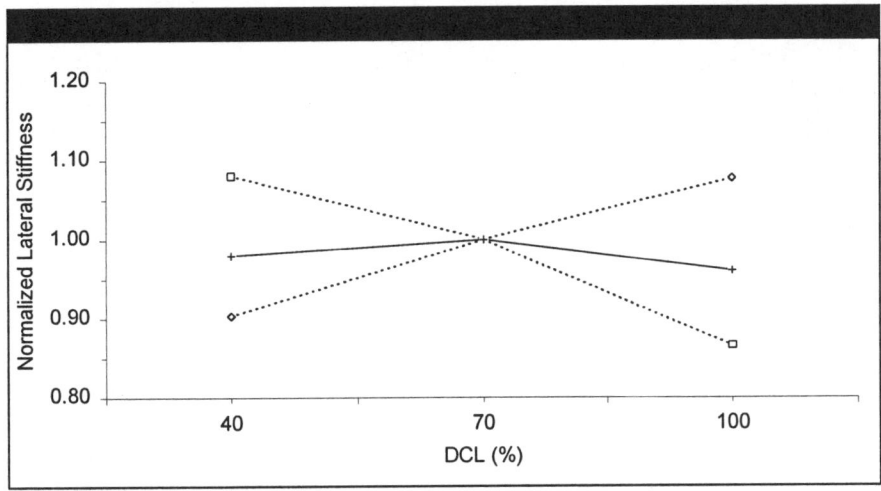

Figure 4.2(b) Compressive Load Dependency: Normalized Stiffness Compressive Load Dependency

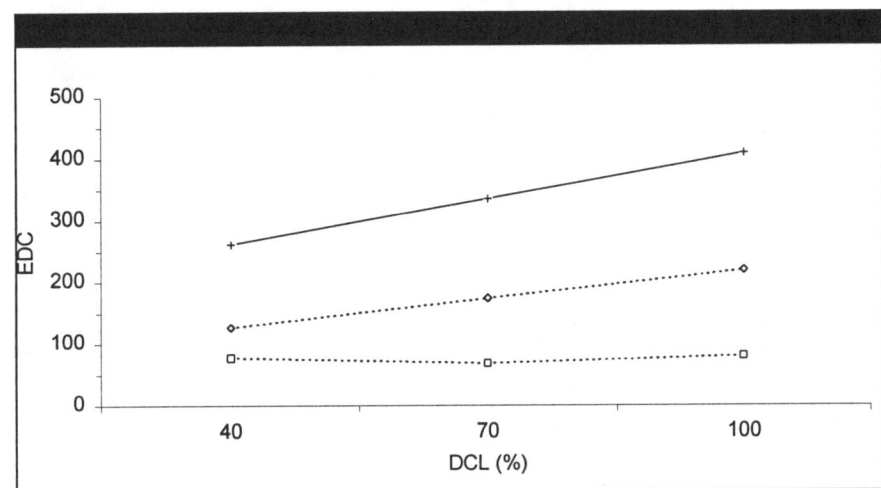

Figure 4.2(c) Compressive Load Dependency: EDC Compressive Load Dependency

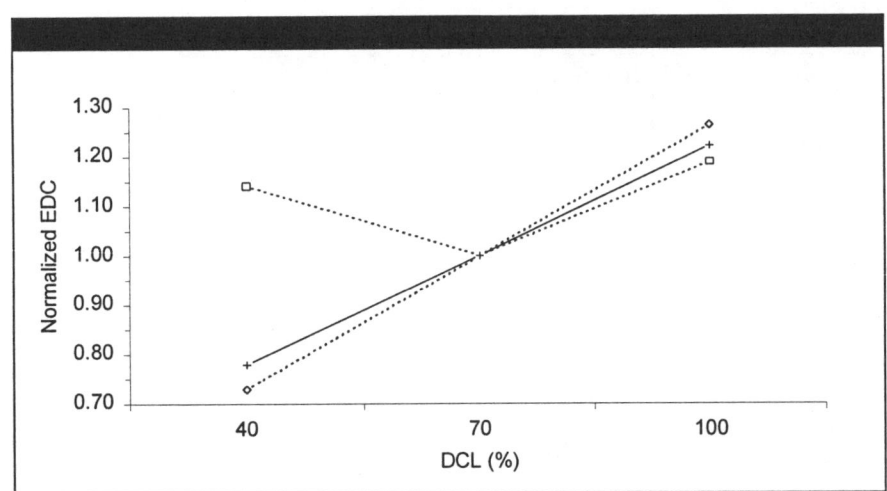

Figure 4.2(d) Compressive Load Dependency: Normalized EDC Compressive Load Dependency

Data Summary, Analysis, and Review

TA # 2 while the stiffness increased with increasing DCL for TA # 3. The stiffness remained nearly constant for TA # 5. Equivalent damping and EDC increased with increasing DCL on TA # 3 and TA #5. However, equivalent damping and EDC went up and down with increasing DCL on TA # 2.

Delaminations between the steel shims/load plates and rubber laminates observed during Test 1 on TAs # 2, #4 and #5 continued to propagate. The delaminations between the top/bottom load plates and the rubber laminates on TA # 5 allowed the bearing to rock during the cyclic testing.

4.3 Test 3 – Frequency Dependent Characterization

Purpose
To determine experimentally dynamic performance characteristics at varying frequencies in the primary direction of operation.

Procedure
With the DCL applied, three fully reversed cycles of the design displacement were applied at frequencies corresponding to periods of 20.0, 5.0, 2.0, 1.0, and 0.5 seconds. The resulting data was used to quantify performance characteristics, specifically, stiffness, damping, and EDC of weight-bearing devices under dynamic loading conditions.

Test Variances
Test variances were necessary while testing Tekton's SRB devices. Each variance was documented and is recorded in this report.

Variance:	The 0.5 second period tests were performed at 50 percent of the design displacement.
Reason:	Testing velocity for the large bearing (TA #5) exceeded the capacity of the test facility for the 0.5 second period at full design displacement.
Resolution:	All isolators were tested at 0.5 DD for consistency (0.5 second period).
Impact:	Performance characteristics for the 0.5 second period cannot be directly compared with other periods.

Table 4.3 Test 3 — Frequency Dependent Characterization (Second Cycle)

Test Article ID	Period (seconds)				
	20.0	5.0	2.0*	1.0	0.5**
Stiffness [kips/in]					
TA #1 (150 kip)	❖	❖	❖	❖	❖
TA #2 (500 kip)	3.4	3.3	3.2	3.7	3.5
TA #3 (500 kip)	5.2	5.5	5.6	5.4	4.7
TA #4 (500 kip)	3.6	3.7	–	3.8	4.0
TA #5 (750 kip)	4.7	4.7	4.8	5.1	6.8
Damping [% Critical]					
TA #1 (150 kip)	❖	❖	❖	❖	❖
TA #2 (500 kip)	8.4	9.1	9.2	9.0	7.3
TA #3 (500 kip)	15.0	14.5	14.3	15.0	17.5
TA #4 (500 kip)	8.3	8.3	–	7.5	5.4
TA #5 (750 kip)	15.5	14.6	14.2	13.6	7.6
EDC [in-kips]					
TA #1 (150 kip)	❖	❖	❖	❖	❖
TA #2 (500 kip)	80.7	81.8	80.7	88.8	18.5
TA #3 (500 kip)	220.6	218.9	219.7	214.9	59.8
TA #4 (500 kip)	83.8	84.0	–	77.5	15.6
TA #5 (750 kip)	448.0	421.2	410.1	408.7	81.0

* Data for 2.0 second period is from Test 2 at 100% DCL.
** Performed at 50% DD.
❖ Test 3 was not performed on TA #1 due to a failure during Test 1.

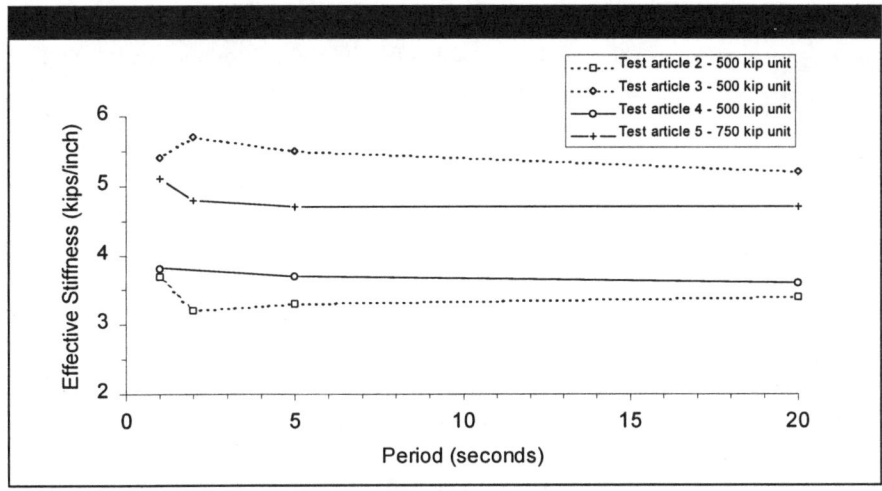

Figure 4.3(a) Frequency Dependency: Stiffness Frequency Dependency

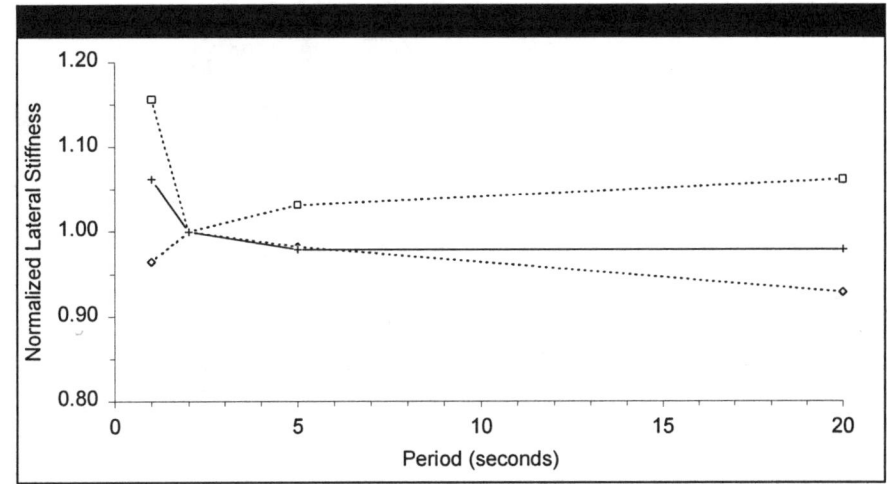

Figure 4.3(b) Frequency Dependency: Normalized Stiffness Frequency Dependency

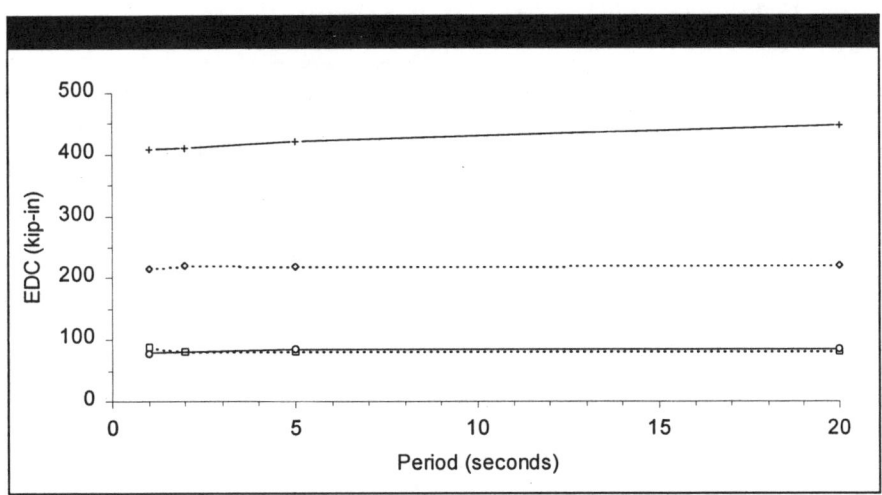

Figure 4.3(c) Frequency Dependency: EDC Frequency Dependency

Data Summary, Analysis, and Review

Data Summary

The results of Test 3, performed to characterize stiffness, EDC, and damping as a function of frequency, are summarized in Table 4.3 and Figures 4.3(a-d).

Test Observations

Frequency dependency trends were not observed during this test. Delaminations between the steel shims/load plates and rubber laminates observed during Test 1 and Test 2 on TAs #2, #4, and #5 continued to propagate. The delaminations between the top/bottom load plates and the rubber laminates on TA # 5 allowed the bearing to rock during the cyclic testing.

4.4 Test 5 – Fatigue and Wear

Purpose

To evaluate the potential seismic performance changes resulting from 10,000 cycles of service movements (temperature and live load fluctuations).

Procedure

The device was subjected to a minimum of 10,000 cycles of simulated displacement representing the movement rating (MR) of the device specified by the manufacturer and applied at a frequency corresponding to a 10.0 second period or other frequency deemed acceptable to the manufacturer. The Evaluation Plan required that the speed of application be at least 4.5 inches per minute. If the component was designed as an isolation device and provides vertical support, i.e., a load bearing isolator, the full DCL was applied. The resulting data were used to quantify damage or degradation arising from the application of high-frequency, small-displacement loading.

Test Variances

No variances were reported.

Data Summary

Deterioration from fatigue and wear was not evident based upon visual inspection.

Test Observations

No specific damage was observed.

4.5 Test 6 – Environmental Aging

Purpose

To verify experimentally seismic performance after exposure to a salt spray environment.

Procedure

The device was exposed in a salt spray chamber for 1,000 hours in accordance with the requirements of ASTM B117. The accelerated aging testing examined the degradation of the device resulting from key environmental factors.

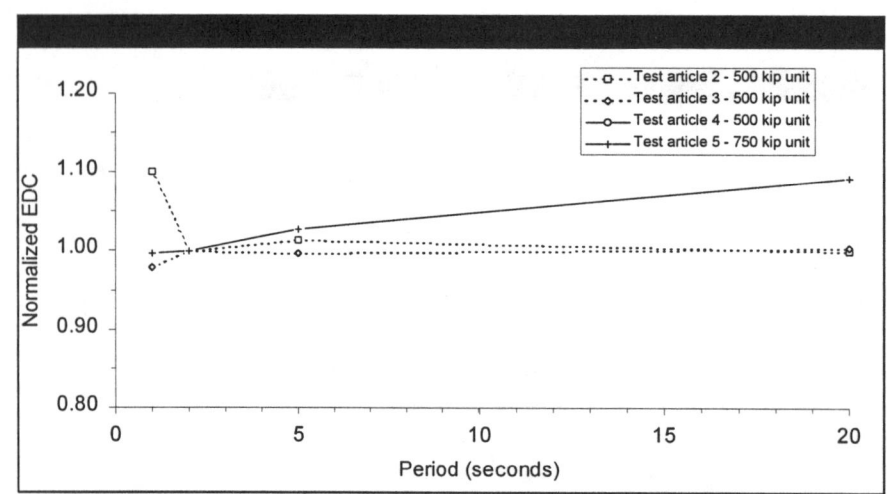

Figure 4.3(d) Frequency Dependency: Normalized EDC Frequency Dependency

Test Variances

No variances were reported.

Data Summary

An outside laboratory performed salt spray conditioning on TA #4 (500 kip unit). The device was returned for testing without the optional warm water rinsing. Test 3, Frequency Dependent Characterization, was then used to assess changes in performance.

The accelerated aging effects of fatigue and wear (Test 5) and salt spray (Test 6) on stiffness, EDC, and damping are summarized in Table 4.4.

Test Observations

The stiffness, damping, and EDC increased after the combined 10,000 cycle and salt spray test.

4.6 Test 7 – Dynamic Performance Characteristics at Temperature Extremes

Purpose

To assess the effects of extreme temperature on the performance characteristics, specifically, stiffness, damping, and EDC.

Procedure

With the full DCL applied (for load bearing isolators), three fully reversed cycles of the design displacement were applied at a frequency corresponding to a 2.0 second period at the upper and lower temperature extremes specified by the

Table 4.4 Tests 5 and 6 – Fatigue and Wear and Environmental Aging (Second Cycle)

Period (Seconds)	Stiffness (kips/in)		Damping (% Critical)		EDC (in-kips)	
	Before**	After	Before**	After	Before**	After
20.0	3.6	3.8	8.3	10.3	83.8	110.4
5.0	3.7	3.9	8.3	9.4	84.0	100.4
1.0	3.8	4.1	7.5	8.6	77.5	96.2
0.5*	4.0	4.7	5.4	6.8	15.6	23.3

* Performed at 50% DD.

** "Before" data is taken from Table 4.3, TA #4.

Table 4.5 Test 7 – Dynamic Performance Characteristics at Temperature Extremes (Second Cycle)

Performance Parameters	Cold Temperature 49 hrs @ -20° F*	Ambient Temperature 70° F	Hot Temperature 24 hrs @ 120° F
Stiffness (kips/in)	6.0 (+88%)*	3.2	3.9 (+22%)*
Damping (% Critical)	13.4 (+46%)*	9.2	6.8 (-26%)*
EDC [in-kips]	230 (+185%)*	80.7	76.7 (-5%)*

* () indicates percent change from ambient temperature test results.

manufacturer. The temperature range of interest for this evaluation program was from -40° F to 120° F. However, if specific devices could not perform at these temperature extremes, manufacturer-specified temperature limits were used.

Further test procedures for handling hot or cold test articles were subsequently developed as follows:

1) The test article and mounting hardware were placed in the heating/cooling unit for the prescribed number of hours;

2) The test article and mounting hardware were installed in the test rig within 75 minutes after being removed from the thermal chamber; and

3) Testing was performed within five minutes after installation was complete.

The resulting data quantified changes in performance that occurred as a result of changes in ambient temperature conditions.

Test Variances

Test variances were necessary while testing Tekton's SRB devices. Each variance was documented and is recorded in this report.

Variance: TA #2 was stored in the thermal chamber for 49 hours before testing, instead of 24 hours.
Reason: Rubber characteristically continues to stiffen with increased exposure to cold temperatures. In order to make the test data more useful to the user, the length of exposure in the thermal chamber was increased for devices manufactured with rubber.
Impact: Additional stiffening was achieved.

Data Summary

Temperature conditioning and testing were performed on TA #2 (500 kip unit). The effects of temperature on stiffness, damping, and EDC are summarized in Table 4.5. The upper and lower temperature limits were specified or approved by Tekton. The manufacturer specified a low temperature operation limit of -20°F.

Test Observations

Stiffness, damping, and EDC increased dramatically with decreasing temperature. Stiffness also increased slightly with

increased temperature; however, damping and EDC decreased with increased temperature. Since nine months elapsed between the ambient temperature test and the extreme temperature test, it is possible that some of the performance changes could be attributed to the effect of aging on the rubber.

4.7 Test 8 – Durability

Purpose

To assess component durability resulting from a moderate number of strong motion cycles.

Procedure

Twenty fully reversed cycles at 100 percent maximum design displacement were applied at a frequency corresponding to a 2.0 second period. The full DCL was applied simultaneously to the weight-bearing isolators. The resulting data was used to assess performance degradation, quantify useable margins, and provide insight into the suitability of the device for applications where a large number of high-level aftershocks can be expected.

Test Variances

Test variances were necessary while testing Tekton's SRB devices. Each variance was documented and is recorded in this report.

Variance: The test was interrupted at the end of the fifth cycle (TA #5).
Reason: The number of test cycles completed was limited by nitrogen left in the accumulator from previous test.
Resolution: The accumulators were recharged and the remainder of the 20 cycles was completed.
Impact: Twenty cycles could not be completed without interruption. The interruption may have affected the total degradation recorded.

Data Summary

In addition to demonstrating the ability to withstand a significant number of design displacement loadings, Test 8 can also be viewed as an extension of the stability test (Test 1).

Table 4.6 Test 8 – Durability

Test Article	F_1(kips)	F_5(kips)	F_{10}(kips)	F_{15}(kips)	F_{20}(kips)
TA #1 (150 kip)	*	*	*	*	*
TA #2 (500 kip)	28	26	25	25	24
TA #3 (500 kip)	44	41	40	39	38
TA #5 (500 kip)**	60	54	–	–	–
**	–	57 (F_6)	52	49	47

* Test 8 was not performed on TA #1 due to failure during Test 1.
** Split Test (five cycles and 15 cycles).
F_i = Peak lateral force during i^{th} cycle.

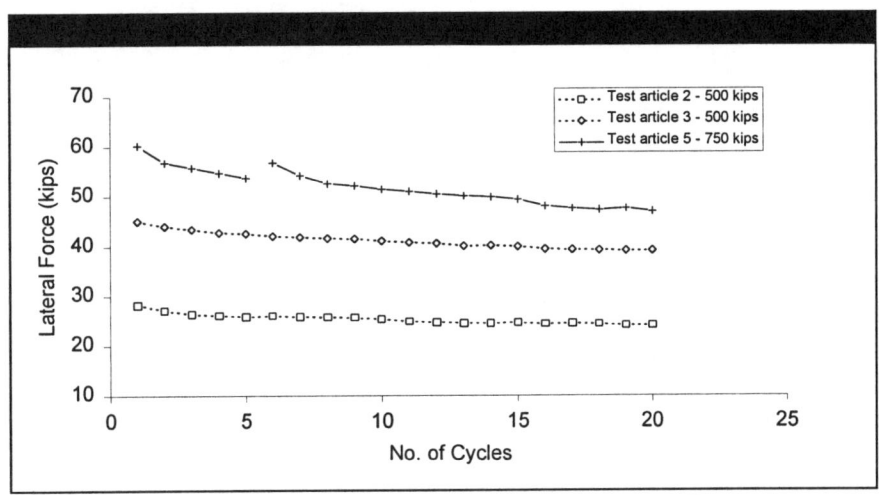

Figure 4.4(a): Force Degradation During Durability Test: Force Degradation

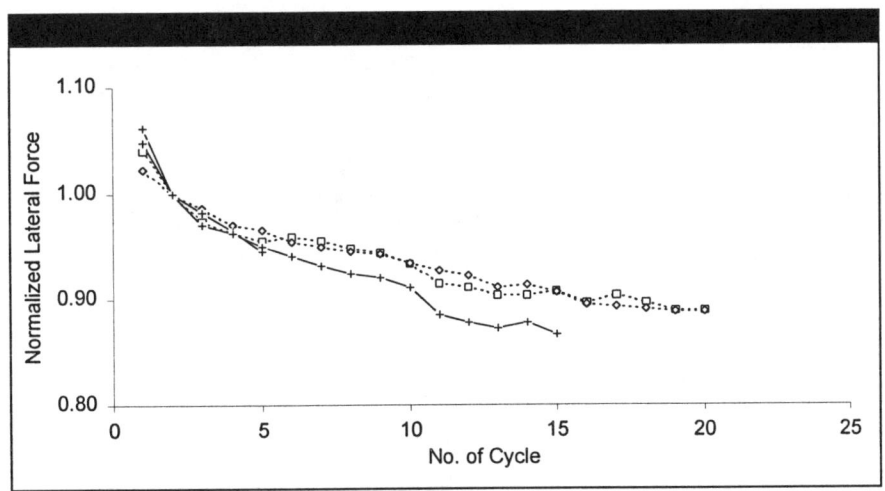

Figure 4.4 (b) Force Degradation During Durability Test: Normalized Force Degradation

Data Summary, Analysis, and Review

Table 4.7 Test 9 – Ultimate Performance

Test Article ID	Design Disp. (DD) (in)	Failure Disp. (FD) (in)	Safety Margin (FD/DD)
TA #1 (150 kip)	6.0	*	*
TA #2 (500 kip)	6.75	14.5	2.15
TA #3 (500 kip)	6.75	16.7	2.47
TA #4 (500 kip)	6.75	13.1	1.94
TA #5 (750 kip)	10.0	**	**

* Failure displacement was not determined due to failure during Test 1.
** Failure displacement was not determined due to limits of test equipment.

Test Observations

TAs #2, #3, and #5 sustained 20 fully reversed cycles at the design displacement. Although TA #2 and TA #3 were the same size, their performance was markedly different. Test data show that TA #3 was 50 percent stiffer than TA #2; however, both test articles exhibited similar rates of degradation. The 750 kip bearing experienced greater and more rapid stiffness degradation than the two smaller bearings (Figure 4.4[b]). The force degradation curve for the 750 kip bearing also indicated that stopping the test on the fifth cycle allowed the bearing to restiffen to nearly pre-test stiffness.

4.8 Test 9 – Ultimate Performance

Purpose

To determine experimentally ultimate displacements and margins of safety.

Procedure

Two test methods were used to complete this evaluation.

Test 9.1: With the DCL applied, the device was loaded laterally at a frequency corresponding to a 2.0 second period with increasing displacement amplitude until "failure" occurred. Each cycle increased the displacement amplitude by 1.1 times the preceding cycle, e.g., $(1.1)^0$xDD, $(1.1)^1$xDD, $(1.1)^2$xDD, $(1.1)^3$xDD ... $(1.1)^n$xDD for the nth cycle. For some devices, physical failure may be an actual "failure," whereas for other devices it was a programmed physical stop or ultimate restraint beyond which the device could not operate.

Test 9.2: If failure did not occur before the cyclic limits of the test machine (± 15 inches) were reached, the device was subjected to a slow rate monotonic loading to 22 inches, the test machine limit.

Table 4.8 Comparison of Predicted and Measured Performance

Test Article Size	Effective Stiffness (kips/in)		Equivalent Damping (% Critical)		EDC (in-kips)	
	Design	Actual*	Design	Actual*	Design	Actual*
150 kip**	3.87	2.4	10	>100	88	630
500 kip	10.5	5.6	8.0	14.3	240	219.7
750 kip	9.47	4.8	12	14.2	714	410.1

* "Actual" values are taken from Table 4.3 for TAs #3 and #5 at 100% DCL for the second cycle and 2.0 second period.
** "Actual" values for TA #1 were calculated from Test 1 results since TA #1 did not undergo Test 2.

The resulting data was used to examine safety margins and provide insight into the stability of the device under large displacement. Failure of an elastomeric device was defined as: 1) the elastomeric material visually/audibly tears or delaminates, or 2) the lateral shear stiffness degrades dramatically.

Test Variances

Test variances were necessary while testing Tekton's SRB devices. Each variance was documented and is recorded in this report.

Variance: TA #5 testing ended at the completion of Test 9.1. The monotonic loading (22 inches) was not performed.

Reason: The expected failure load for TA #5 was over 150 kips and could damage testing equipment.

Impact: No data on ultimate displacement is available for TA #5.

Data Summary

The displacements to failure and margins of safety are summarized in Table 4.7. Design displacements were based on 100 percent, 75 percent, and 83 percent shear strain for the 150 kip, 500 kip, and 750 kip isolators, respectively.

Test Observations

TA #2 failed the increasing amplitude harmonic cycling test (Test 9.1) at a lateral displacement of 14.5 inches when the rubber laminate delaminated completely from the upper load plate.

TA #3 completed the increasing amplitude harmonic cycling test (Test 9.1) at a maximum displacement of 15 inches. The outer protective cover became detached from the top and bottom load plates. TA #3 failed during the increasing amplitude push test (Test 9.2) at a displacement of 16.7 inches when the lateral load dropped quickly. Sand escaped from the steel pin cavities.

TA #4 failed at a lateral displacement of 13.1 inches during the increasing amplitude harmonic cycling test (Test 9.1). The delamination between the top and bottom load plates propagated with each successive stroke.

TA # 5 completed the increasing amplitude harmonic testing (Test 9.1) up to the maximum displacement of 15.0 inches. The push test (Test 9.2) was not performed because of concerns that the test rig carriage bearing might become overloaded.

4.9 Test Predictions

Predictions submitted by the manufacturer and the actual measured performance are provided in Table 4.8. Design values were based on information provided by the manufacturer. Their methods for calculation of design values may have been different from the methods used to interpret the test data for this report. For other periods, including static test (20 second period), please refer to Table 4.3

CHAPTER 5

Summary

In March 1996, HITEC published *Guidelines for the Testing of Seismic Isolation and Energy Dissipating Devices*, an Evaluation Plan, under the guidance of a Technical Evaluation Panel. This Plan described test methods used to evaluate the performance of several different types of isolators and dampers manufactured by a number of companies. The Evaluation Plan specified design parameters for the isolators and dampers, tests to be performed, and testing instructions. Full scale, dynamic tests were specified to measure performance benchmarks, compressive load dependency, frequency dependency, fatigue and wear effects, effects of environmental aging, dynamic performance at extreme temperatures, durability, and ultimate performance. Testing was subsequently completed for 11 technologies from 10 companies.

Tekton Inc. manufactures and sells steel rubber isolation bearings (SRB). The bearings consist of alternating layers of rubber and steel shims. As part of this evaluation, Tekton manufactured five bearings for testing at the ETEC facility. The bearings consisted of three sizes with design axial-load capabilities of 150 kip, 500 kip (three bearings), and 750 kip. According to the test plan, the bearings originally had design displacements of six inches, nine inches, and 12 inches respectively. However, after early test results associated with the 150 kip device, Tekton revised their design displacements from nine inches to 6.75 inches and from 12 inches to 10 inches for the 500 kip and 750 kip devices, respectively.

Based on the test results, the following observations were made:

- During the initial performance benchmark test, the 150 kip bearing exhibited unstable performance and an excessive compressive set. Furthermore, extensive distortion of the outer cover was noted. This was classified as a failure and no further tests were performed on the 150 kip bearing. The manufacturer modified the remaining devices and reduced the design displacements before benchmark testing resumed. The rate of degradation for these devices was rapid and unstable. Delaminations were observed on the first and third 500 kip devices and the 750 kip device.

- The bearings were tested with varying DCL. Some of the performance measurements changed with DCL; however, no trends were obvious. The delaminations observed during the benchmark testing continued to propagate during the DCL testing. The rubber laminate on the 750 kip bearing allowed it to rock.

- The bearings were not frequency dependent. However, during the frequency testing the delaminations continued to propagate.

- No deterioration from fatigue and wear was evident from visual inspection of the third 500 kip device. The same test article was then exposed to a salt spray environment. Further testing showed that the stiffness, damping, and EDC generally increased after the combined fatigue and salt spray exposures.

- Stiffness, damping, and EDC increased dramatically with decreasing temperature (-20F). Stiffness also increased slightly with increased temperature; however, damping and EDC decreased with increased temperature. The results of the temperature testing may have been affected by a delay of nine months between the ambient temperature testing and the extreme temperature testing.

- The second and third 500 kip bearings and the 750 kip bearing were subjected to 20 fully reversed cycles at the design displacement. The two 500 kip bearings had a markedly different performance. The second 500 kip bearing was 50 percent stiffer than the second 500 kip bearing; however, both bearings exhibited similar degradation rates. The 750 kip bearing experience greater and more rapid stiffness degradation than both the 500 kip devices.

- The 500 kip bearings were loaded to failure with either an increasing amplitude harmonic cyclic loading or a monotonic push loading. The 750 kip device completed the increasing amplitude harmonic loading. The three 500 kip bearings had displacement safety margins of 2.15, 2.47, and 1.94.

Appendix A

Selected Test Plots

This appendix contains the characteristic plots of selected test runs. Additional plots listed in Appendix C are available upon request from HITEC.

Over 80 plots were generated for each isolator tested. A select number of representative plots are provided in the body of this report for TA's #1 (150 kip), #3 (500 kip), and #5 (750 kip). The test numbers given on these plots are internal numbers for the testing facility and are not meaningful within the context of this report.

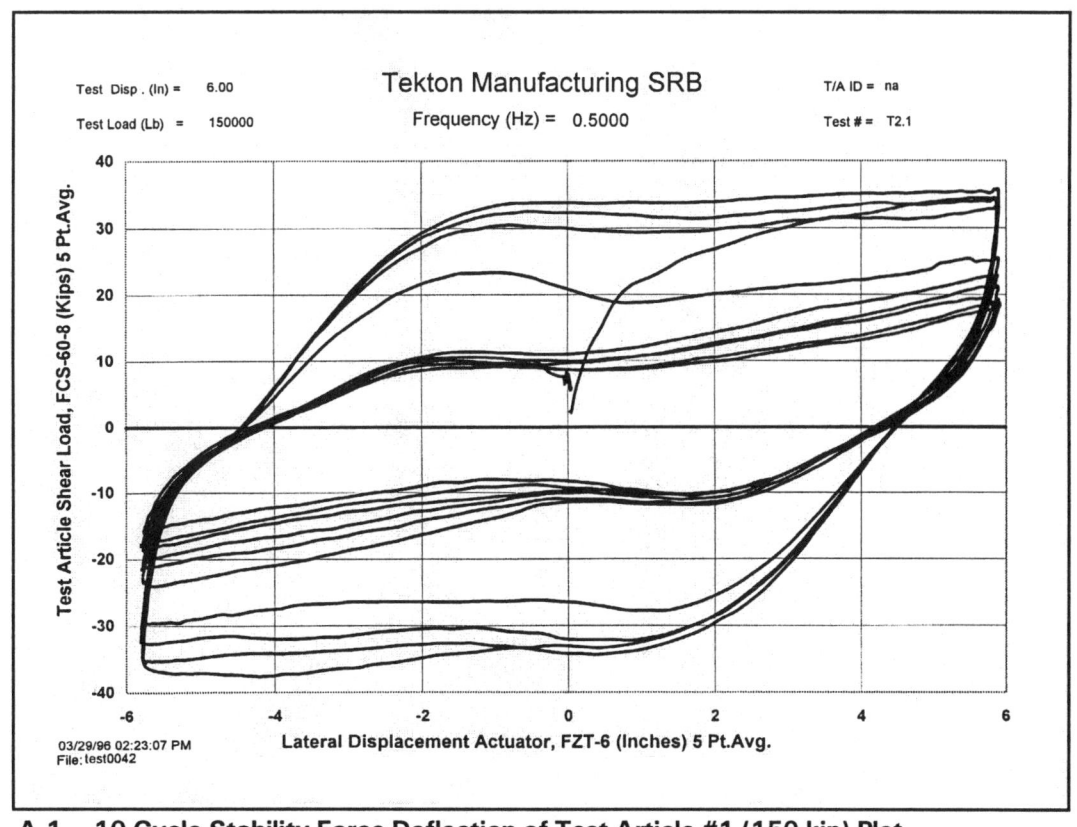

A.1 10 Cycle Stability Force Deflection of Test Article #1 (150 kip) Plot

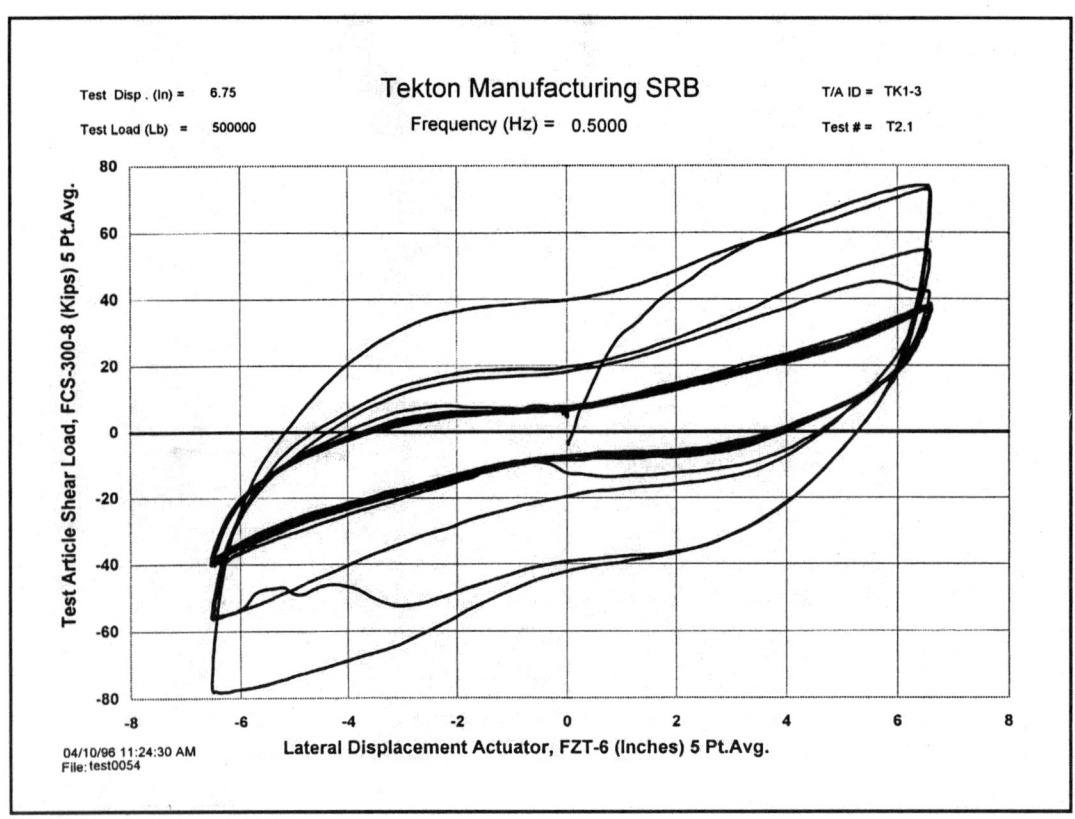

A.2   10 Cycle Stability Force Deflection of Test Article #3 (500 kip) Plot

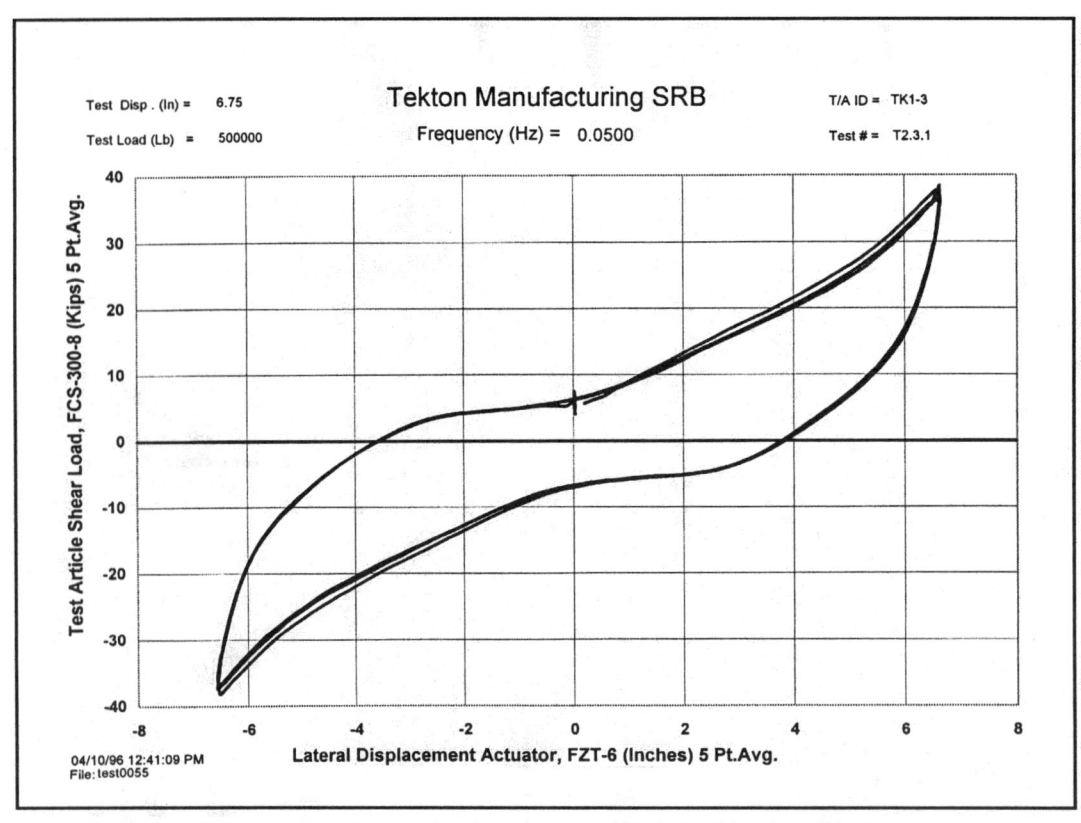

A.3   0.05 Hz Force Deflection Response Test Article #3 (500 kip) Plot

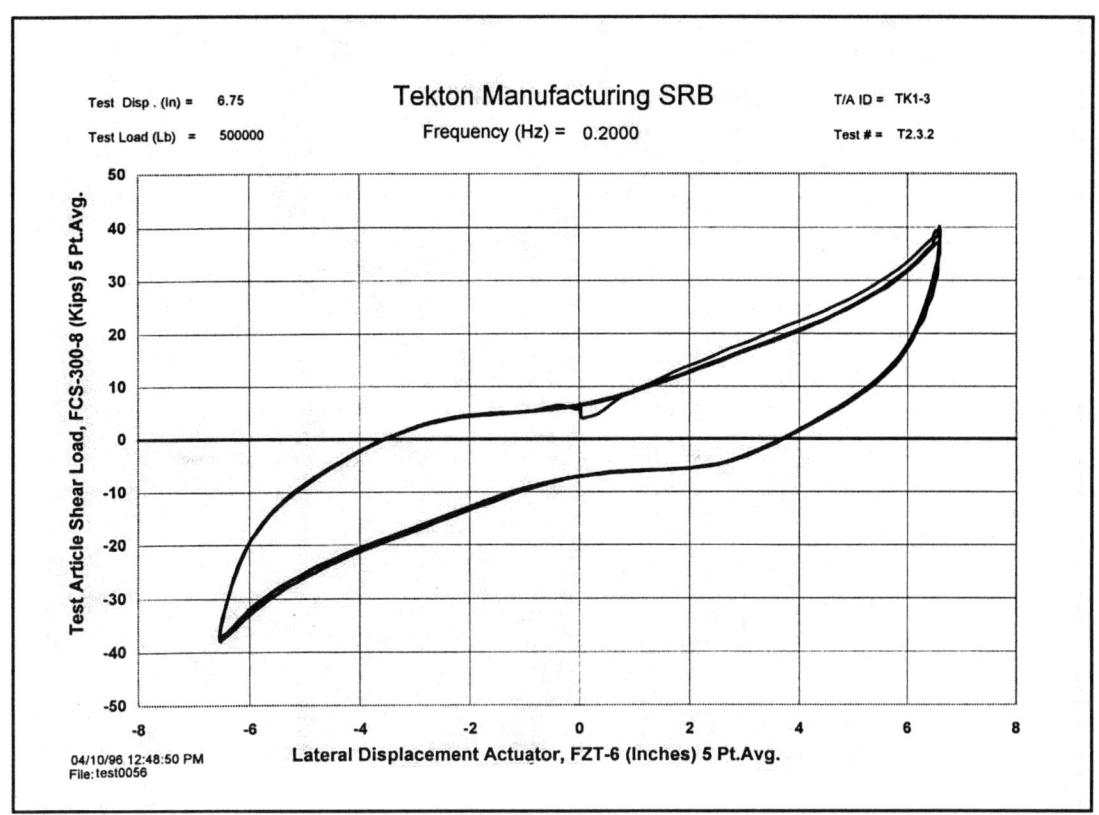

A.4   0.20 Hz Force Deflection Response Test Article #3 (500 kip) Plot

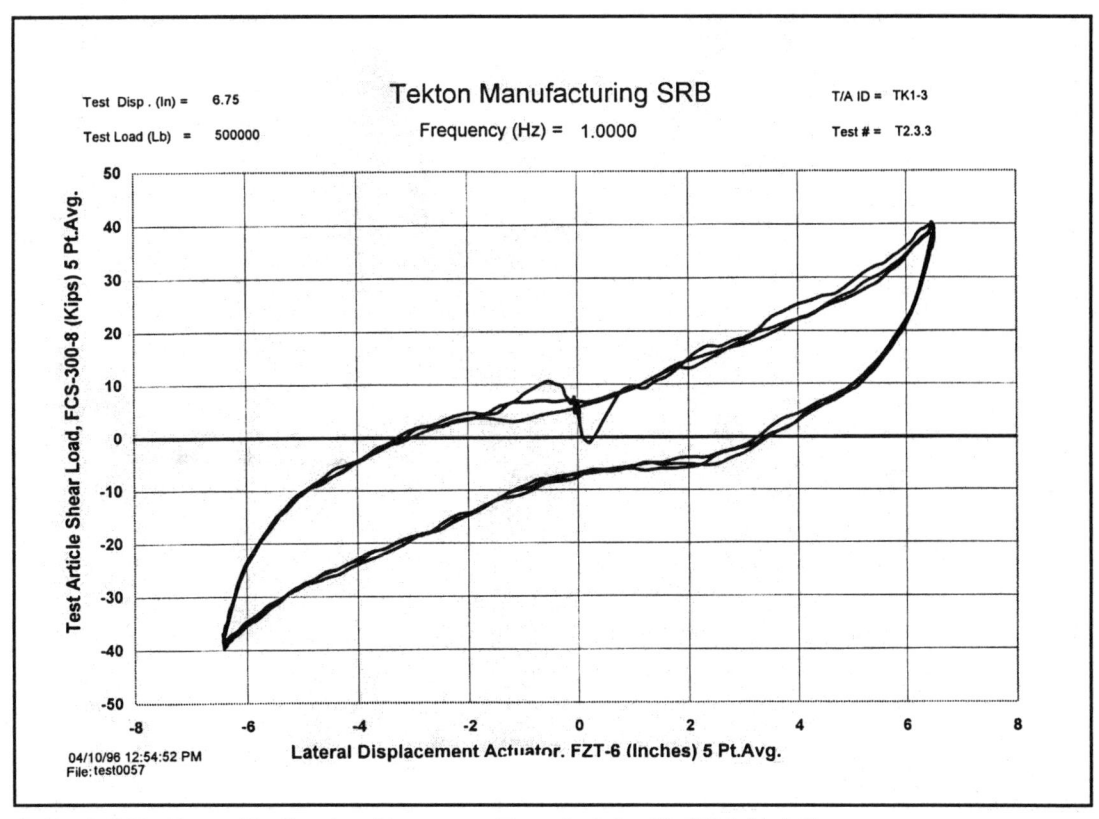

A.5   1.0 Hz Force Deflection Response Test Article #3 (500 kip) Plot

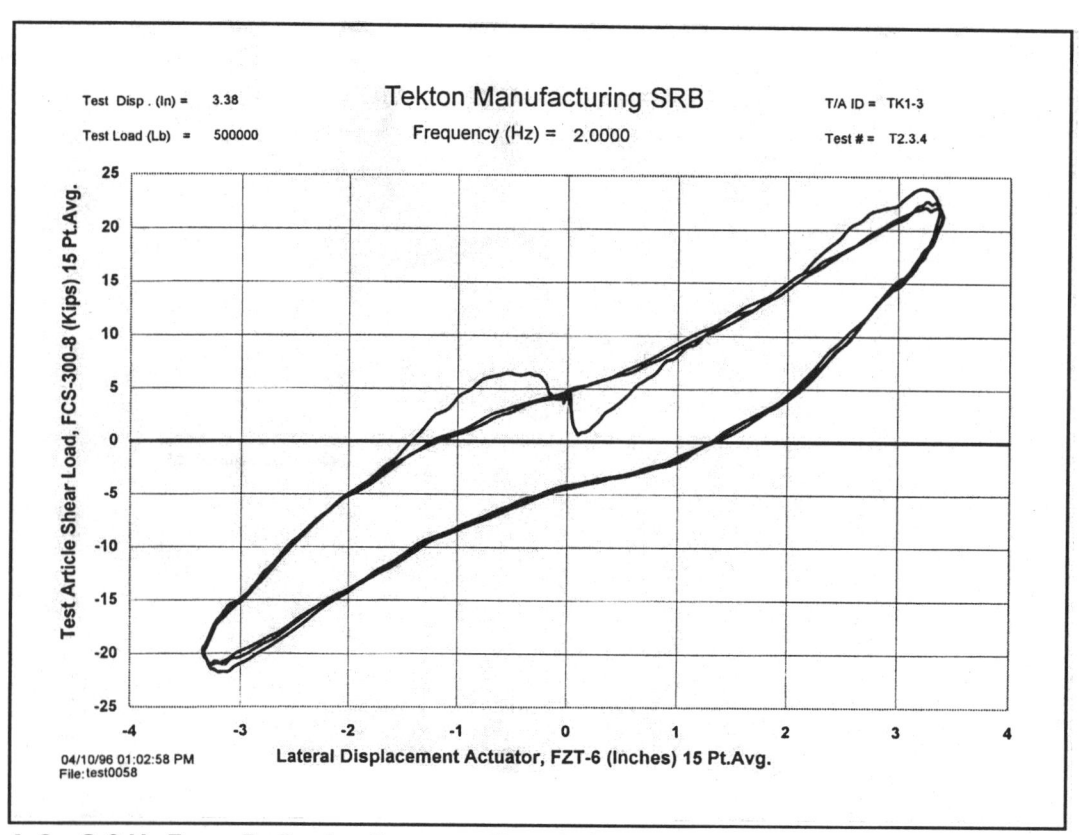

A.6 2.0 Hz Force Deflection Response Test Article #3 (500 kip) Plot

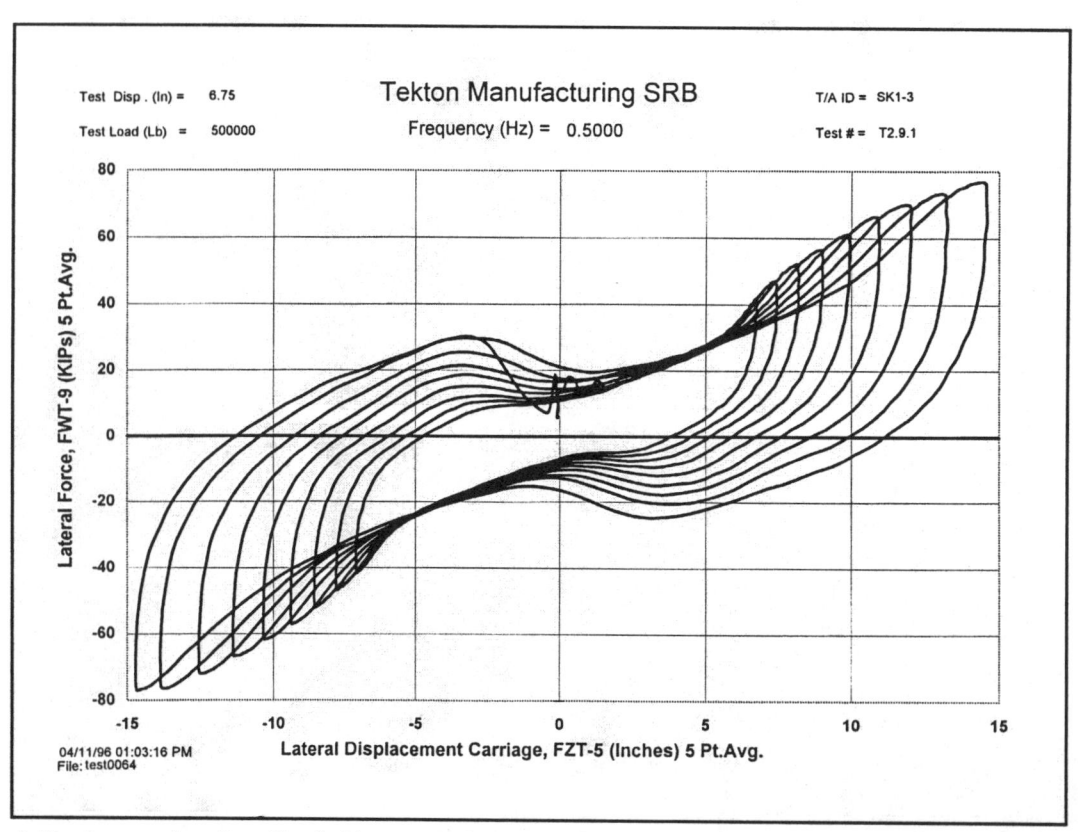

A.7 Increasing Amplitude Harmonic Loading (Test 9.1) Test Article #3 (500 kip) Plot

HITEC: Tekton® Steel Rubber Bearings

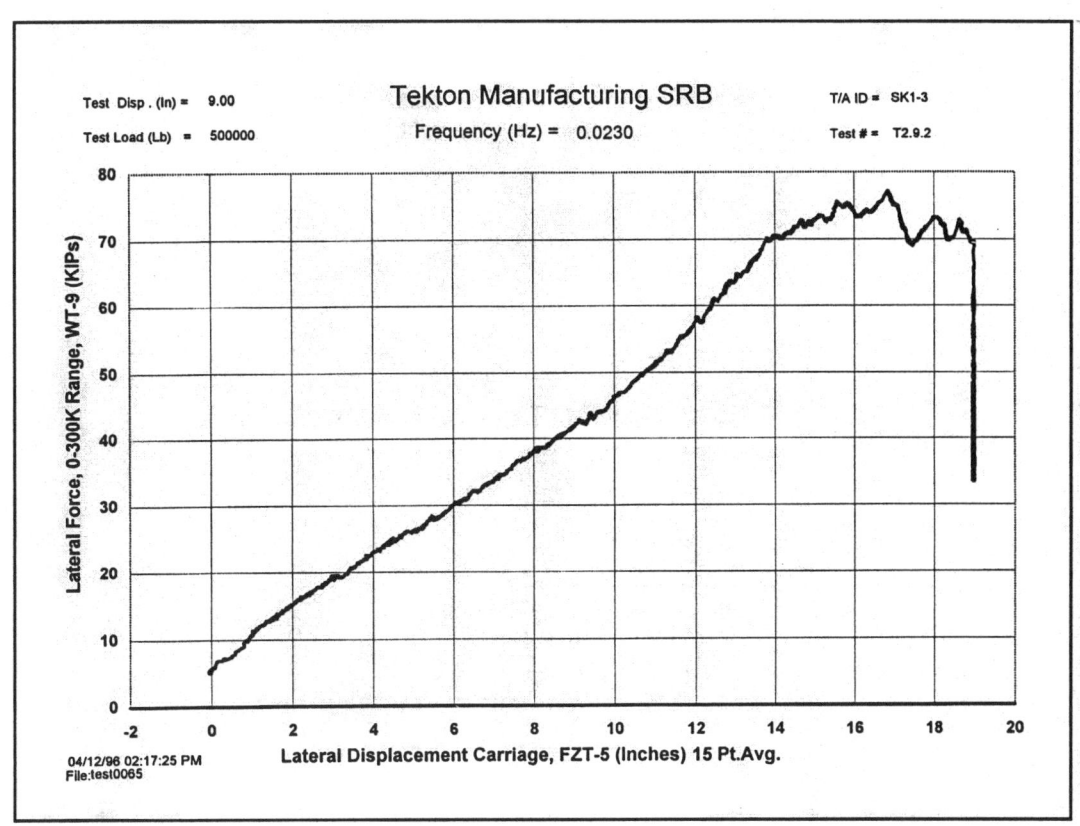

A.8   Increasing Amplitude Push Test (Test 9.2) Test Article #3 (500 kip) Plot

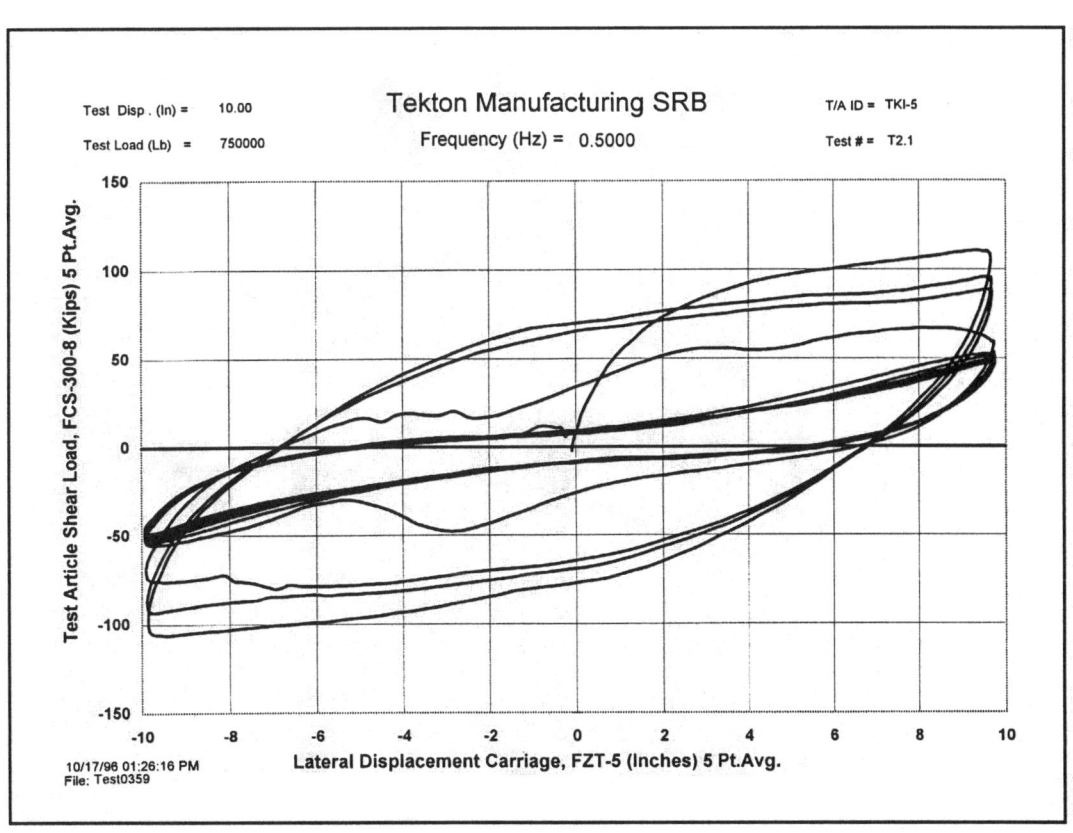

A.9   10 Cycle Stability Force Deflection of Test Article #5 (750 kip) Plot

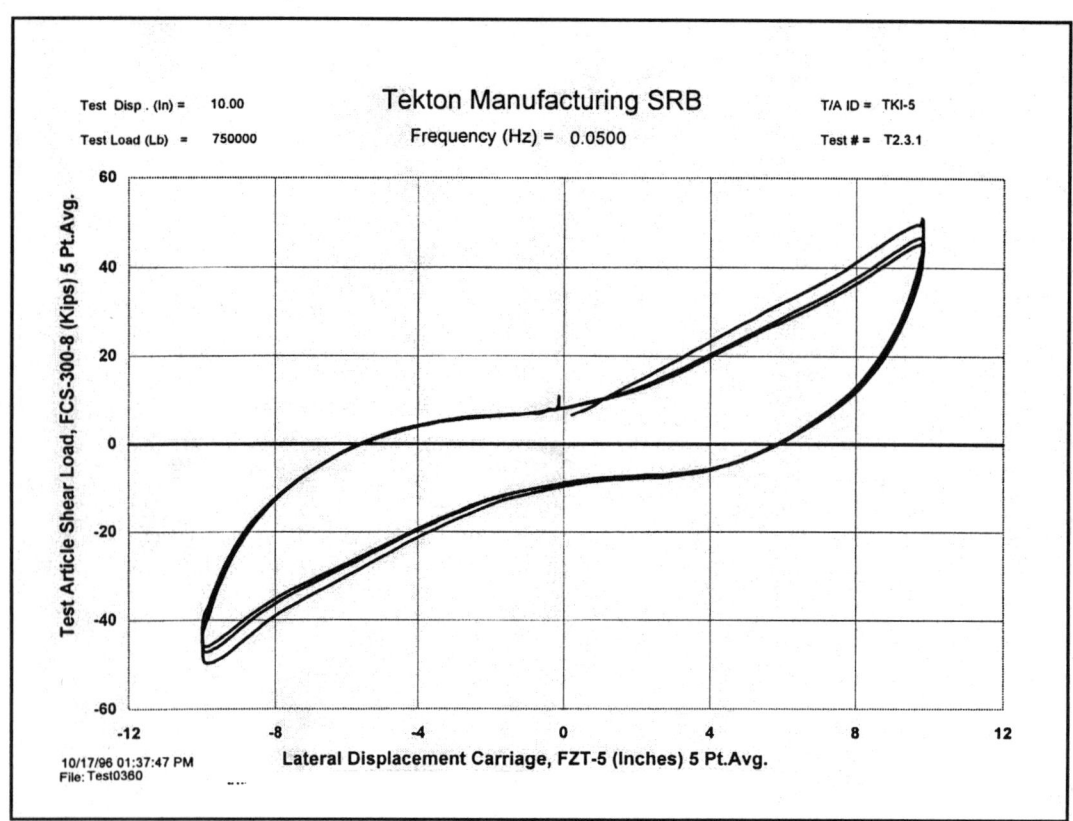

A.10   0.05 Hz Force Deflection Response Test Article #5 (750 kip) Plot

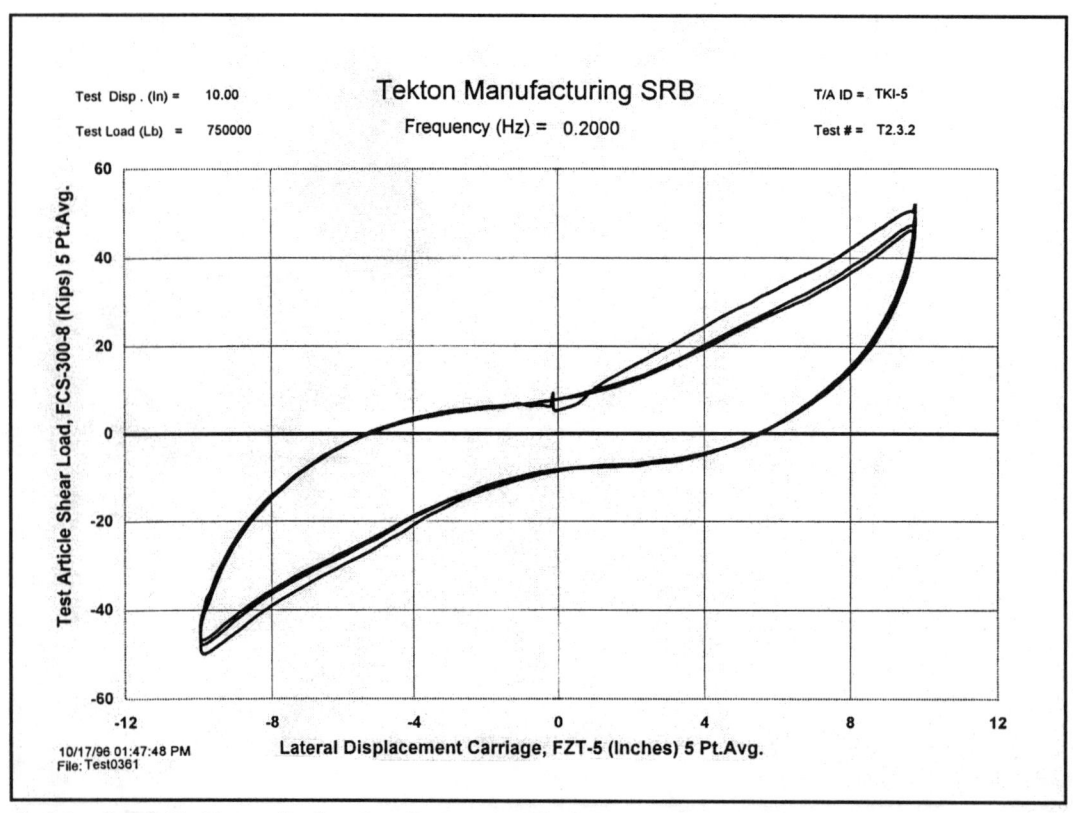

A.11   0.20 Hz Force Deflection Response Test Article #5 (750 kip) Plot

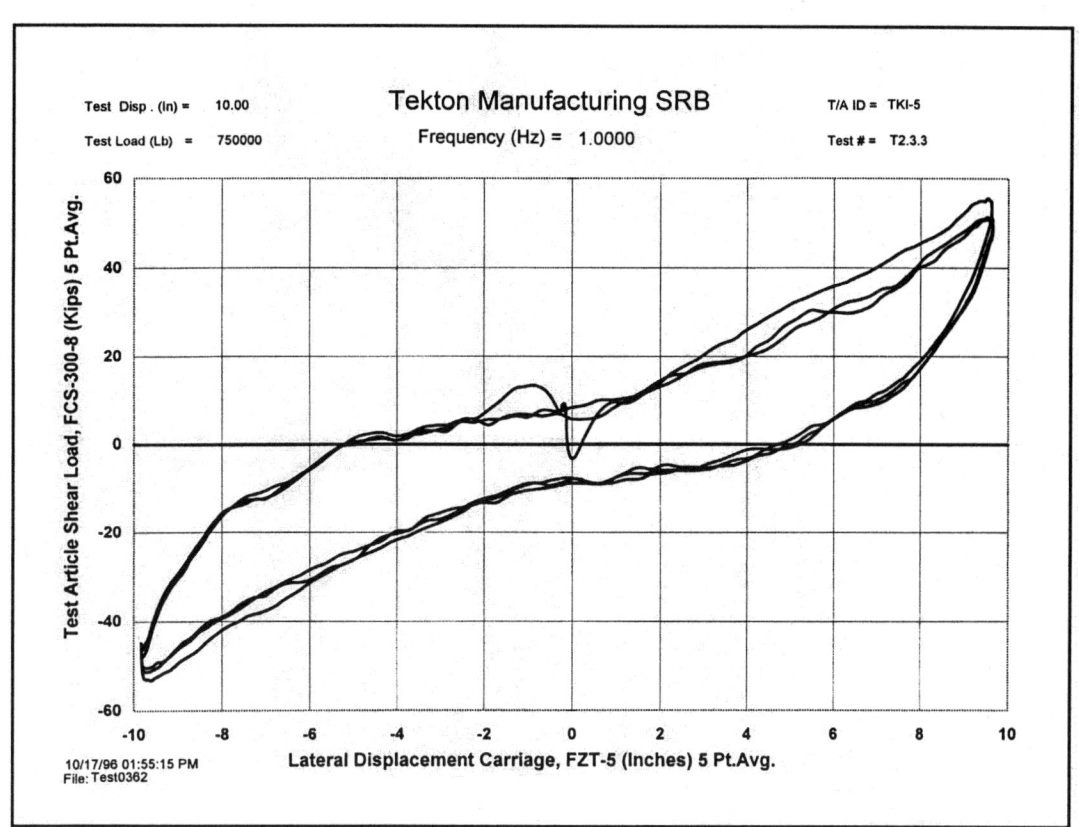

Test Disp . (In) =   10.00
Test Load (Lb) =   750000

Tekton Manufacturing SRB

Frequency (Hz) = 1.0000

T/A ID =   TKI-5

Test # =   T2.3.3

10/17/96 01:55:15 PM
File: Test0362

A.12   1.0 Hz Force Deflection Response Test Article #5 (750 kip) Plot

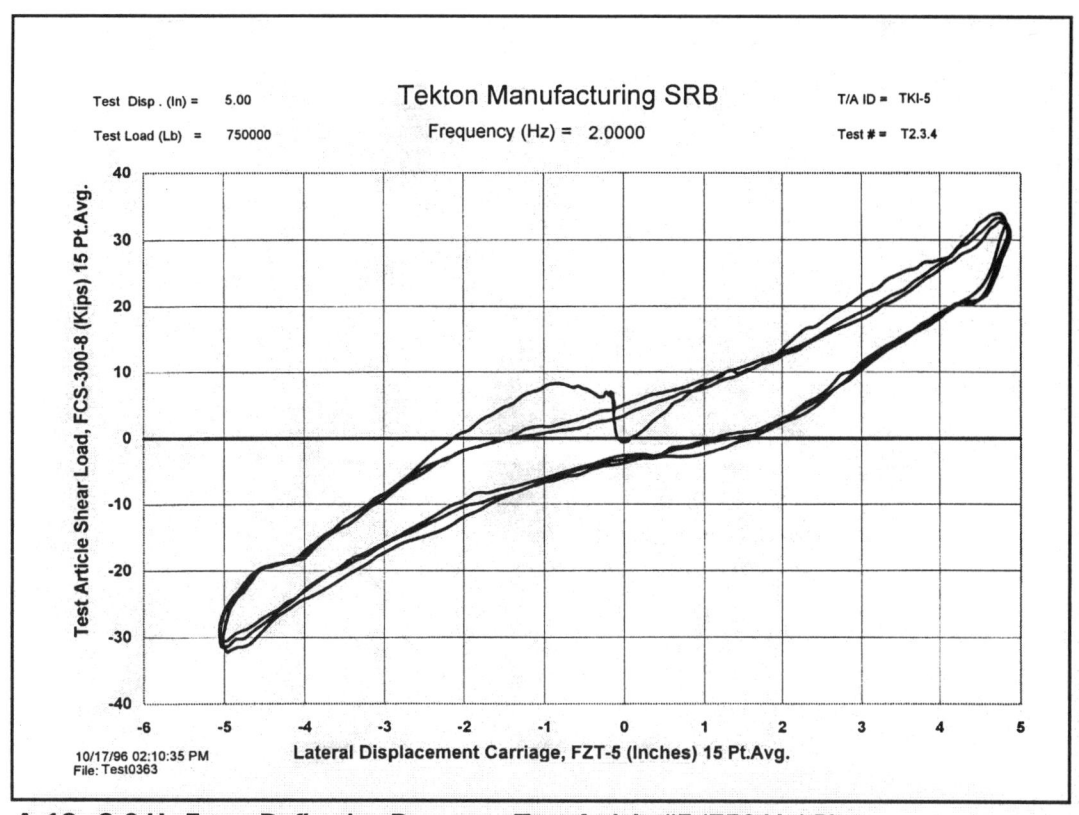

Test Disp . (In) =   5.00
Test Load (Lb) =   750000

Tekton Manufacturing SRB

Frequency (Hz) = 2.0000

T/A ID =   TKI-5

Test # =   T2.3.4

10/17/96 02:10:35 PM
File: Test0363

A.13   2.0 Hz Force Deflection Response Test Article #5 (750 kip) Plot

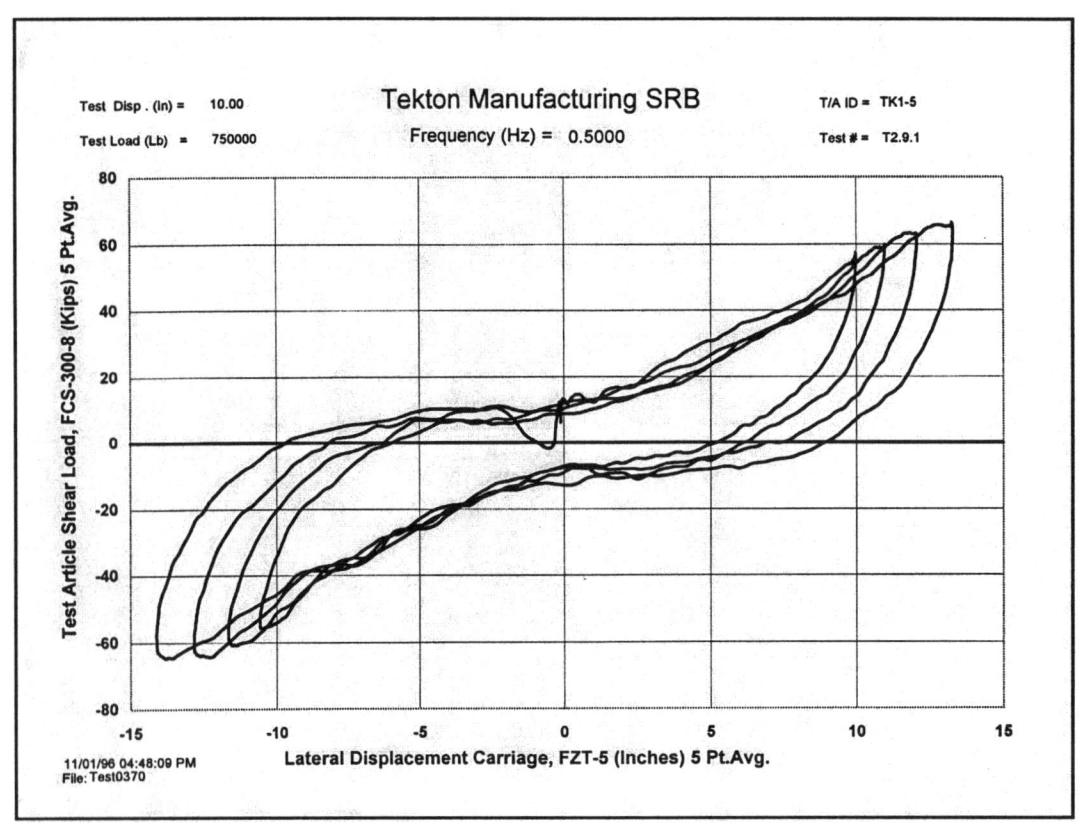

A.14 Increasing Amplitude Harmonic Loading (Test 9.1) Test Article #5 (750 kip) Plot

Appendix B

Reported Stiffness and Damping Properties

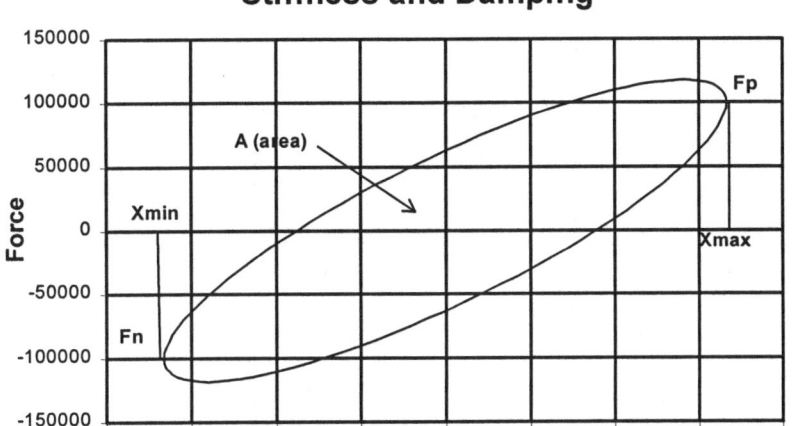

Stiffness and Damping

$$K_{eff} = (F_p - F_n)/(X_{max} - X_{min})$$

$$X_{max}, \text{and } X_{min} = \text{Zero Velocity Displacements}$$

$$\text{Damping ratio } (\beta) = [1/(2\pi)] * \text{Area}/(K_{eff} * X_o^2)$$

$$X_o = \text{Design Displacement}$$

$$\{ X_o = (X_{max} - X_{min})/2 \}$$

Appendix C

Available Data Plots

The following plots are available from HITEC upon request:

Test 1 – Performance Benchmark (DCL and DD at 0.5 Hz)

Lateral Force vs. Lateral Displacement
Lateral Force vs. Time
Average Vertical Displacement vs. Time
Lateral Displacement vs. Time
Total Compressive Force vs. Time
Total Compressive Force vs. Lateral Displacement
Average Vertical Displacement vs. Lateral Displacement

Test 2 – Compressive Load Dependent Characterization (DD at 0.4 DCL, 0.7 DCL and 1.0 DCL, 0 Degree and 90 Degree Rotations)

Lateral Force vs. Lateral Displacement
Lateral Force vs. Time
Lateral Displacement vs. Time
Average Vertical Displacement vs. Time
Total Compressive Force vs. Time
Total Compressive Force vs. Lateral Displacement
Average Vertical Displacement vs. Lateral Displacement

Test 3 – Frequency Dependent Characterization (DD and DCL at 0.05, 0.20, and 1.0 Hz, and 0.5 DD at 2.0 Hz)

Lateral Force vs. Lateral Displacement
Lateral Force vs. Time
Lateral Displacement vs. Time
Average Vertical Displacement vs. Time
Total Compressive Force vs. Time
Total Compressive Force vs. Lateral Displacement
Average Vertical Displacement vs. Lateral Displacement

Test 5 – Fatigue and Wear (10,000 cycles at MR and DCL) – TA #4 Only

No Plots Recorded

Test 6 – Environmental Aging

No Plots Recorded

Test 7 – Dynamic Performance at Temperature Extremes (DD and DCL at 0.5 Hz, at Hot and Cold Tests)

Lateral Force vs. Lateral Displacement
Lateral Force vs. Time
Lateral Displacement vs. Time
Average Vertical Displacement vs. Time
Total Compressive Force vs. Time
Total Compressive Force vs. Lateral Displacement
Average Vertical Displacement vs. Lateral Displacement

Test 8 – Durability (DD and DCL at 0.5 Hz)

Lateral Force vs. Lateral Displacement
Lateral Force vs. Time
Lateral Displacement vs. Time
Average Vertical Displacement vs. Time
Total Compressive Force vs. Time
Total Compressive Force vs. Lateral Displacement
Average Vertical Displacement vs. Lateral Displacement

Test 9.1 – Ultimate Performance (DCL, Increasing Displacement and 0.5 Hz)

Lateral Force vs. Lateral Displacement
Lateral Force vs. Time

Test 9.2 – Ultimate Performance (DCL, Monotonic Increasing Displacement)

Lateral Force vs. Lateral Displacement
Lateral Force vs. Time

Glossary

AASHTO. American Association of State Highway and Transportation Officials.

ASCE. American Society of Civil Engineers.

ASTM. American Society for Testing and Materials.

Caltrans. California Department of Transportation.

CERF. Civil Engineering Research Foundation.

Damping. The ability to dissipate energy.

Design Displacement (DD). The maximum lateral displacement under seismic loading.

Design Compressive Load (DCL). The maximum design vertical load (dead load, live load, overturning, etc.).

Equivalent Damping. Value of equivalent viscous damping corresponding to the energy dissipated during cyclic response at the design displacement of the isolator.

Effective Stiffness. See Appendix B.

Energy Dissipation per Cycle (EDC). Area under force deflection hysterisis loop.

Energy Dissipater. A device that is used to dissipate energy by friction or viscous flow.

ETEC. Energy Technology Engineering Center.

FHWA. Federal Highway Administration.

HITEC. Highway Innovative Technology Evaluation Center.

Isolator. A device that is used to isolate a structure from seismic ground motion.

Motion-Controlled. Specified motion that an isolator or energy dissipater is forced to follow in a test.

Movement Rating (MR). The small displacement range (lateral) of the device due to temperature and live load fluctuations (excluding earthquakes).

Number of Shake-Down Cycles. Number of fully reversed motion-controlled cycles required for an isolator or energy dissipater to repeat its performance or stabilize its cyclic response.

TA(s). Test article(s).

TRB. Transportation Research Board.